Technologies Enabling Future Mobile Connectivity and Sensing

RIVER PUBLISHERS SERIES IN COMMUNICATIONS AND NETWORKING

The "River Publishers Series in Communications and Networking" is a series of comprehensive academic and professional books which focus on communication and network systems. Topics range from the theory and use of systems involving all terminals, computers, and information processors to wired and wireless networks and network layouts, protocols, architectures, and implementations. Also covered are developments stemming from new market demands in systems, products, and technologies such as personal communications services, multimedia systems, enterprise networks, and optical communications.

The series includes research monographs, edited volumes, handbooks and textbooks, providing professionals, researchers, educators, and advanced students in the field with an invaluable insight into the latest research and developments.

Topics included in this series include:-

- Communication theory
- Multimedia systems
- Network architecture
- Optical communications
- Personal communication services
- Telecoms networks
- Wi-Fi network protocols

For a list of other books in this series, visit www.riverpublishers.com

Technologies Enabling Future Mobile Connectivity and Sensing

Editors

Björn Debaillie
imec, Belgium

François Brunier
Soitec, France

Dominique Morche
CEA-Leti, France

Erkan Nevzat Isa
Fraunhofer EMFT, Germany

Jan Craninckx
imec, Belgium

River Publishers

Routledge
Taylor & Francis Group
NEW YORK AND LONDON

Published 2023 by River Publishers
River Publishers
Alsbjergvej 10, 9260 Gistrup, Denmark
www.riverpublishers.com

Distributed exclusively by Routledge
605 Third Avenue, New York, NY 10017, USA
4 Park Square, Milton Park, Abingdon, Oxon OX14 4RN

Technologies Enabling Future Mobile Connectivity and Sensing by Björn Debaillie, François Brunier, Dominique Morche, Erkan Nevzat Isa, Jan Craninckx

Routledge is an imprint of the Taylor & Francis Group, an informa business

ISBN 9788770040747 (hardback)
ISBN 9781003812265 (online)
ISBN 9781032633039 (master ebook)

Dedication

"The wireless revolution is not just about convenience, but about the power to connect people, places, and ideas in ways never before possible."

Guglielmo Marconi

"The speed of connectivity determines the speed of innovation."

Reed Hastings

"Nature is our kindest friend and best critic in experimental science if we only allow her intimations to fall unbiased on our minds."

Michael Faraday

Acknowledgement

The editors would like to thank all the contributors for their support in the planning and preparation of this book. The recommendations and opinions expressed in the book are those of the editors, authors, and contributors and do not necessarily represent those of any organizations, employers, or companies.

Björn Debaillie
François Brunier
Dominique Morche
Erkan Nevzat Isa
Jan Craninckx

Contents

Preface xi

Editors' Biography xvii

List of Contributors xix

List of Figures xxiii

List of Tables xxxi

1. RF technology roadmap for 5G and 6G RF front end systems 1
Yvan Morandini

2. From beyond 5G to sub-THz era 11
François Brunier

3. A low power 5G access point targeting airplane cabin connectivity 19
Fredrik Tillman, Pallavi Paliwal, and Daniel Eckerbert

4. Digital beamforming transceiver design in 22nm FD-SOI technology for 39 GHz 5G access 31
Jérôme Prouvée, Giovanni Mangraviti, Björn Debaillie, Piet Wambacq, David Borggreve, Radu Ciocoveanu, Henrik Fredriksson, Pallavi Paliwal, Fredrik Tillman, Hasip Terlemez, Burak Dündar, Vincent Pinon, Fredric Hasbani, Alexandre Ferret, Cedric Dehos, Abdelaziz Hamani, Baudouin Martineau, Dominique Morche

5. Disruptive TRX design for D-band **57**
Jose Luis González-Jiménez, Alexandre Siligaris, Abdelaziz Hamani, Cédric Dehos, Fabrice Chaix, Pierre Courouve, Guillaume Robe, Jean Baptiste David, Francesco Foglia Manzillo, Antonio Clemente, Nicolas Cassiau

6. Packaging technologies and challenges towards 5G integration of mm-wave components and Silicon ICs **71**
Tekfouy Lim, Michael Kaiser, Mattis Obst, Tanja Braun, Marius van Dijk, Johannes Jaeschke, Lars Böttcher, Kavin Senthil Murugesan, Uwe Maass, Ivan Ndip

7. Energy efficient beam control for 5G antennas **87**
Damian Duraj, Luiza Leszkowska, Weronika Kalista, Kamil Trzebiatowski, Lukasz Kulas, Krzysztof Nyka

8. Recent progress on SOI CMOS power amplifiers for mobile and WiFi applications **103**
Alexandre Giry, Ayssar Serhan, Ali Alshakoush, Pascal Reynier

9. V2X RF front end module assistance circuits: observation receiver, DPD, and supply modulator **115**
Emre Ulusoy, Bahadır Özkan, Fatih Maden, Furkan Barin, Ercem Yeşil, Adem Eren, Dursun Baran, Tufan Coşkun Karalar, Ahmet Tekin, Ertan Zencir

10. Impact of high-resistivity substrate on RF and mm-wave performance of 22 nm FD-SOI devices and circuits **127**
Martin Rack, Lucas Nyssens, Massinissa Nabet, Dimitri Lederer, Jean-Pierre Raskin

11. Detection of human targets using a MIMO FMCW radar with slow-time DC-Value suppression **143**
Keivan Alirezazad, Linus Maurer

Appendix

12. Some visions towards 6G cellular systems **151**
Aarno Pärssinen

13. 6G communications and localization: an overview of technologies opportunities and challenges **153**
Didier Belot

14. Heterogeneous integration for complex mm-wave transceivers **155**
Piet Wambacq

15. D-Band noise characterization and modelling in advanced FDSOI devices **157**
Thomas Kämpfe

Index **159**

Preface

Technologies Enabling Future Mobile Connectivity and Sensing

In today's connected world, the demand for mobile communications and instant access to information, anytime and anywhere has drastically changed the electronics landscape, both consumer and industrial. Novel 5G and 6G systems will enable connectivity in all forms between humans, devices, machines and any objects. They will provide virtually ubiquitous, ultra-high bandwidth and low latency network access to individual users, as well as to all objects benefiting from being connected. They will be the "eyes and ears" of artificial intelligence systems as they will provide real-time data collection and analysis. Such diversity calls for a new paradigm in terms of flexibility, not only related to performance, but also in terms of scalability and cost.

5G and 6G communication systems imply a major stake of sovereignty and autonomy for the communication sector and digital infrastructures of the future. All products related to IoT, traffic and health care, supported by connectivity, will benefit the citizens in their daily lives to improve everything from business to private affairs. Together, this will influence society as much as smart phones did in the recent past. It is all about communication and connectivity.

In this respect it is most essential, for Europe, to have full technological sovereignty with an aspect of trusted manufacturing, without dependency on foreign solutions that potentially comprise hidden agendas, e.g. embedded pernicious functionalities in commercial equipment. As such, the BEYOND5 project has built a completely European supply chain for radio-frequency electronics, enabling new RF domains for sensing, communication, 5G radio infrastructure and beyond. It has gathered the most significant European actors covering the entire value chain from materials, semiconductor technologies, designs and components up to the systems. A key pillar was the support and development of the most advanced SOI technologies manufactured in Europe, namely RFSOI and FDSOI, and their use in systems with simultaneous demands for large-scale integration, low power consumption, cost competitiveness and high reliability.

Moving forward, into higher frequency bands above 100 GHz for 6G, more disruptive technologies, using heterogeneous integration of CMOS, SOI, and III/V components such as GaN or InP and advanced packaging techniques, will play an even more dominant role in realizing the objectives of ubiquitous, ultra-high bandwidth and low latency networks.

This book offers complete coverage of the topics presented at the International Workshop on "Technologies Enabling Future Mobile Connectivity and Sensing" in Lisbon, Portugal, 11 September 2023, as part of the ESSCIRC/ESSDERC 2023 European Solid-state Circuits and Devices Conference.

This book provides an overview of the latest research results in this field. It is based on the close collaboration in the BEYOND5 project, extended with vision and roadmap insights by European experts leading 6G development. Through articles and abstracts, a combined view of experts and practitioners representing academia, research and industry in the field of wireless communication systems is given. They cover the topics of RF and digital SOI technology development for 5G and 6G, device and substrate characterization, packaging technology and the realization of full systems, including power amplifiers, linearization techniques, beamforming transceivers, access points and radar detection.

This book is a valuable resource for researchers, designers, developers, academics, post-graduate students and practitioners seeking recent research results on 5G and 6G technology. It combines the latest developments on both SOI and heterogeneous technology, and their use in full application systems. As such, it offers insight into important technological trends for novel and future wireless connectivity systems. The book is structured into 11 articles and 4 abstract descriptions. A brief introduction of each article and abstract distributed over three different sessions is discussed in the following paragraphs.

The first set of abstracts and papers focus on the vision and roadmap of mobile applications and their technology choices.

- *Aarno Pärssinen**: "Some visions towards 6G cellular systems" presents the world's first 6G research and testbed program and discusses some visions, roadmaps and challenges to realize the next generation cellular systems.

---

* The presentation abstracts are grouped together in the one common chapter following the full paper chapters.

- *Didier Belot**: "6G communications and localization: an overview of technologies opportunities and challenges" presents the key challenges and opportunities in terms of solid-state RF circuits to enable 6G to utilize the higher frequency bands towards THz.
- *Yvan Morandini*: "RF technology roadmap for 5G and 6G RF front end systems" addresses the considerable challenge for the RF front and designs and their implementation when moving towards 6G by in presenting an RF technology roadmap including engineering substrates.
- *François Brunier*: "From beyond 5G to the sub-THz era" introduces the key ambitions and activities of the BEYOND5 project to establish a complete and reliable European SOI supply chain and to accelerate the co-innovation pace and market adoption, and also looks to alternative technologies and ecosystem building to support the development of systems operating at sub-THz frequency bands.

The second set of papers and abstract focus on different system design and integration approaches for next generation communication.

- *Fredrik Tillman, Pallavi Paliwal, Daniel Eckerbert*: "A low power 5G access point targeting airplane cabin connectivity" discusses opportunities for inflight 5G communication, and how a possible in-cabin 5G MIMO radio could be specified and implemented.
- *Jérôme Prouvée, Giovanni Mangraviti, Björn Debaillie, Piet Wambacq, David Borggreve, Radu Ciocoveanu, Henrik Fredriksson, Pallavi Paliwal, Fredrik Tillman, Hasip Terlemez, Burak Dündar, Vincent Pinon, Fredric Hasbani, Alexandre Ferret, Cedric Dehos, Abdelaziz Hamani, Baudouin Martineau, Dominique Morche*: "Digital beamforming transceiver design in 22nm FD-SOI technology for 39 GHz 5G access" describes a fully integrated transceiver FDSOI design operating at 39 GHz, addressing the power and silicon footprint challenges on a high throughput digital beamforming transceiver system.
- *Jose Luis González-Jiménez, Alexandre Siligaris, Abdelaziz Hamani, Cédric Dehos, Fabrice Chaix, Pierre Courouve, Guillaume Robe, Jean Baptiste David, Francesco Foglia Manzillo, Antonio Clemente, Nicolas Cassiau*: "Disruptive TRX design for D-band" describes an architectu-

* The presentation abstracts are grouped together in the one common chapter following the full paper chapters.

re and RFSOI design for broadband wireless transceivers operating on D-band with the aim of covering tens of GHz of band over the air with a limited bandwidth per channel at the baseband.

- *Piet Wambacq**: "Heterogeneous integration for complex mm-wave transceivers" presents opportunities at technology, integration and packaging levels to sustain the evolution in mobile communications to provide ever more bandwidth in an efficient and commercially viable way.
- *Tekfouy Lim, Michael Kaiser, Mattis Obst, Tanja Braun, Marius van Dijk, Johannes Jaeschke, Lars Böttcher, Kavin Senthil Murugesan, Uwe Maass, Ivan Ndip*: "Packaging technologies and challenges towards 5G integration of mm-wave components and Silicon ICs" elaborates on the packaging technology options to support the high-frequency and high-power requirements of 5G digital beamforming systems and presents an interposer design which addresses different system challenges.
- *Damian Duraj, Luiza Leszkowska, Weronika Kalista, Kamil Trzebiatowski, Lukasz Kulas, Krzysztof Nyka*: "Energy efficient beam control for 5G antennas" describes an overview of different concepts of energy-efficient antenna systems that enhance the connectivity performance of 5G applications and different designs for 5.9 GHz and 39 GHz frequency bands are proposed to improve the capabilities of 5G MIMO systems and reduce energy consumption.

The third set of papers and abstract discuss different component and technology platforms which can enable emerging applications.

- *Alexandre Giry, Ayssar Serhan, Ali Alshakoush, Pascal Reynier*: "Recent progress on SOI CMOS power amplifiers for mobile and WiFi applications" gives a comprehensive overview of recent CMOS SOI PA solutions targeting high-performance mobile and WiFi applications as this technology appears to be an attractive choice for PA integration.
- *Emre Ulusoy, Bahadır Özkan, Fatih Maden, Furkan Barin, Ercem Yeşil, Adem Eren, Dursun Baran, Tufan Coşkun Karalar, Ahmet Tekin, Ertan Zencir*: "V2X RF front end module assistance circuits: observation receiver, DPD, and supply modulator" examines the improved linearity and efficiency of an RF front-end SoC designed in CMOS SOI to meet V2X requirements while avoiding using III–V materials.

* The presentation abstracts are grouped together in the one common chapter following the full paper chapters.

- *Martin Rack, Lucas Nyssens, Massinissa Nabet, Dimitri Lederer, Jean-Pierre Raskin*: "Impact of high-resistivity substrate on RF and mm-wave performance of 22 nm FD-SOI devices and circuits" gives an overview of the increase in performance levels of key mm-wave passive and circuit blocks, and demonstrates that the substrate impacts the main figures of merits in the 20–60 GHz range.
- *Thomas Kämpfe**: "D-Band noise characterization and modelling in advanced FDSOI devices" presents insights into the analysis and the modelling of the high-frequency noise in advanced 22-nm FDSOI CMOS devices operating up to 170 GHz.
- *Keivan Alirezazad, Linus Maurer*: "Detection of human targets using a MIMO FMCW radar with slow-time DC-Value suppression" employs advanced 77 GHz MIMO FMCW radars in combination with an advanced data processing algorithm to detect human presence through their chest movements.

* The presentation abstracts are grouped together in the one common chapter following the full paper chapters.

Editors' Biography

1st Editor: Björn Debaillie, imec, Belgium.

Björn Debaillie (Senior, IEEE) leads imec's collaborative R&D program on cutting-edge IoT technologies, covering high speed communications, high resolution sensing, and neuromorphic computing. As a seasoned researcher & manager, he is responsible for strategic collaborations and partnerships, innovation management, and public funding policies as well as the operational management and coordination across imec's collaborative programs and projects. Björn Debaillie coordinates public funded projects and seeds new initiatives. He holds patents, received awards and authored books and international papers published in various journals and conference proceedings.

2nd Editor: François Brunier, France.

François Brunier (Member, IEEE) graduated as physics and electronics Engineer from Centrale-Supelec in 1997. From 1998 to 2002, he worked as device integration engineer for embedded DRAM products in STMicroelectronics Crolles. In 2002, he joined Soitec as head of advanced characterization laboratory. From 2009 to 2011, as a product manager, he led the RF-SOI and power SOI product development and offering. Since 2012, as a partnership program manager, he is in charge of European collaborative KDT programs, IPCEI and public relations.

3rd Editor: Dominique Morche, CEA-Leti, France.

Dominique Morche (Member, IEEE) received the M.Sc. degree in engineering from the Ecole Nationale Supérieure d'Electricité et de Radioelectricité de Bordeaux, Bordeaux, France, in 1990, and the Ph.D. degree in electronics from the Institut National Polytechnique de Grenoble, Grenoble, France, in

1994. His Ph.D. mainly focuses on sigma–delta AnalogTo-Digital Converter (ADC). From 1994 to 2001, he was with France Telecom, Meylan, France, as a Research Engineer. He has been involved in the architecture and design of analog circuits for telecom applications. He is currently with the Commissariat à l'énergie Atomique-Leti Minatec, Grenoble, where he is a Research Director. His current field of research is in the specification and design of RF architecture for UWB, mmwave, and the IoT systems.

4rd Editor: Erkan Nevzat Isa, Fraunhofer EMFT, Germany.

Erkan Nevzat Isa (Member, IEEE) is with Fraunhofer EMFT since April'13, founding the IC Design group at this center. He holds PhD'12 in Microelectronics from EPFL, MS'06 in Microelectronics from TU Hamburg-Harburg, BS'05 in Computer Engineering from Istanbul Technical University (ITU), and BS'03 in Electronics and Comm. Eng. from ITU. From 2007 to 2010 he was with CEA-Leti, Grenoble, France, working on the design and optimization of A/D converters in deep sub-micron technologies. Between 2010 and 2013 he worked for Fujitsu Semiconductor Europe GmbH in Munich as an analog and mixed-signal design engineer, contributing to the key mixed-signal IPs for automotive SoC products. Dr. Isa serves as Group Manager and have been coordinating collaborative programs for Fraunhofer, e.g. SERENE-IoT (PENTA), THINGS2DO (ENIAC), ADMONT (ECSEL), WAYTOGO FAST (ECSEL), REFERENCE (ECSEL), OCEAN12 (ECSEL), BEYOND5 (ECSEL). Dr. Isa had served as Technical Program Chair and General Co-Chair for IEEE NEWCAS 2018 and NEWCAS 2019, respectively.

5th Editor: Jan Craninckx, imec, Belgium.

Jan Craninckx (Fellow, IEEE) obtained his Ph.D. degree in microelectronics from the KULeuven in 1997 and was with Alcatel Microelectronics as a senior RF engineer until 2002. He then joined IMEC (Leuven, Belgium), where he currently is IMEC fellow for RF, mmwave, analog and mixed signal circuit design. Dr. Craninckx has authored and co-authored more than 200 papers, book chapters and patents, was Editor-in-Chief of the IEEE JSSC, and is an IEEE fellow.

List of Contributors

Alirezazad, Keivan, *Universität der Bundeswehr München, Germany*

Baran, Dursun, *TÜBİTAK, Turkey*

Barin, Furkan, *TÜBİTAK, Turkey, Istanbul Technical University, Turkey*

Belot, Didier, *STMicroelectronics, France*

Borggreve, David, *Fraunhofer EMFT, Germany*

Böttcher, Lars, *Fraunhofer IZM, Germany*

Braun, Tanja, *Fraunhofer IZM, Germany*

Brunier, François, *Soitec, France*

Cassiau, Nicolas, *CEA-Leti, France*

Chaix, Fabrice, *CEA-Leti, France*

Ciocoveanu, Radu, *Fraunhofer EMFT, Germany*

Clemente, Antonio, *CEA-Leti, France*

Courouve, Pierre, *CEA-Leti, France*

David, Jean Baptiste, *CEA-Leti, France*

Debaillie, Björn, *imec, Belgium*

Dehos, Cedric, *CEA-Leti, France*

Dündar, Burak, *MKR, Turkey*

Duraj, Damian, *Gdansk University of Technology, Poland*

Eckerbert, Daniel, *Ericsson, Sweden*

Eren, Adem, *TÜBİTAK, Turkey, Boğaziçi University, Turkey*

Ferret, Alexandre, *ASYGN, France*

Foglia Manzillo, Francesco, *CEA-Leti, France*

Fredriksson, Henrik, *Ericsson, Sweden*

Giry, Alexandre, *CEA-Leti, France*

González-Jiménez, Jose Luis, *CEA-Leti, France*

Hamani, Abdelaziz, *CEA-Leti, France*

Hasbani, Fredric, *ASYGN, France*

Jaeschke, Johannes, *Fraunhofer IZM, Germany*

Kaiser, Michael, *Fraunhofer IZM, Germany*

Kalista, Weronika, *Gdansk University of Technology, Poland*

Kämpfe, Thomas, *Fraunhofer IPMS, Germany*

Karalar, Tufan Coşkun, *TÜBİTAK, Turkey, Istanbul Technical University, Turkey*

Kulas, Lukasz, *Gdansk University of Technology, Poland*

Lederer, Dimitri, *Université catholique de Louvain, Belgium*

Leszkowska, Luiza, *Gdansk University of Technology, Poland*

Lim, Tekfouy, *Fraunhofer IZM, Germany*

Maass, Uwe, *Fraunhofer IZM, Germany*

Maden, Fatih, *TÜBİTAK, Turkey, Istanbul Technical University, Turkey*

Mangraviti, Giovanni, *imec, Belgium*

Martineau, Baudouin, *CEA-Leti, France*

Maurer, Linus, *Universität der Bundeswehr München, Germany*

Morandini, Yvan, *Soitec, France*

Morche, Dominique, *CEA-Leti, France*

Murugesan, Kavin Senthil, *Fraunhofer IZM, Germany, Technical University of Berlin, Germany*

Nabet, Massinissa, *Université catholique de Louvain, Belgium*

Ndip, Ivan, *Fraunhofer IZM, Germany*

Nyka, Krzysztof, *Gdansk University of Technology, Poland*

Nyssens, Lucas, *Université catholique de Louvain, Belgium*

Obst, Mattis, *Fraunhofer IZM, Germany*

Özkan, Bahadır, *TÜBİTAK, Turkey, Istanbul Technical University, Turkey*

Paliwal, Pallavi, *Ericsson, Sweden*

Pärssinen, Aarno, *University of Oulu, Finland*

Pinon, Vincent, *ASYGN, France*

Prouvée, Jérôme, *CEA-Leti, France*

Rack, Martin, *Université catholique de Louvain, Belgium*

Raskin, Jean-Pierre, *Université catholique de Louvain, Belgium*

Reynier, Pascal, *CEA-Leti, France*

Robe, Guillaume, *CEA-Leti, France*

Serhan, Ayssar, *CEA-Leti, France*

Siligaris, Alexandre, *CEA-Leti, France*

Tekin, Ahmet, *Boğaziçi University, Turkey*

Terlemez, Hasip, *MKR, Turkey*

Tillman, Fredrik, *Ericsson, Sweden*

Trzebiatowski, Kamil, *Gdansk University of Technology, Poland*

Ulusoy, Emre, *TÜBİTAK, Turkey, Istanbul Technical University, Turkey*

van Dijk, Marius, *Fraunhofer IZM, Germany*

Wambacq, Piet, *imec, Belgium*

Yeşil, Ercem, *University of California, USA*

Zencir, Ertan, *TÜBİTAK, Turkey, University of Turkish Aeronautical Association, Turkey*

List of Figures

Chapter 1

Figure 1 5G and WiFi spectrum........2
Figure 2 RFFE module content for a 5G mobile handset........2
Figure 3 4G and 5G carrier aggregations........3
Figure 4 RFFE risk interference N12 710 MHz and N66 2130 MHz bands example........4
Figure 5 Third harmonic as a function of the output power comparing HR-SOI with three trap-rich RF SOI flavors with resistivity range from 1 kOhm.cm to 20 kOhm cm at 900 MHz........5
Figure 6 Trap-rich RF-SOI digital noise and comparison with HR-SOI without trap-rich........5
Figure 7 Piezo-on-insulator (POI) versus alternative RF filter technologies benchmarking........6
Figure 8 mmWave RFFE integration options........7
Figure 9 Epitaxy GaN layer........8
Figure 10 Load-pull measurement of a HEMT device with InAlN epitaxy barrier........8

Chapter 2

Figure 1 Schematic illustration of the Smart Cut™ technology used to manufacture SOI substrates........12
Figure 2 Smart Cut™ technology, a source of innovation for a reliable, sustainable, less dependent value chain........12
Figure 3 FD-SOI, an integrated platform selected by worldwide key players........13
Figure 4 KDT project BEYOND5 consortium........13
Figure 5 KDT project BEYOND5 value chain........14
Figure 6 Radio frequency range trend........15
Figure 7 Smart Cut™ of InP........15
Figure 8 Move2THz value chain........16

Chapter 3
Figure 1 Mobile subscriptions by technology (billion)..................20
Figure 2 Global mobile network data traffic (exabyte per month)..................20
Figure 3 Multiple access point installation to achieve needed cabin capacity..................22
Figure 4 Radio interference scenario..................23
Figure 5 Signal levels present at the access point antenna ports.....23
Figure 6 RFIC chip layout showing the four transceivers and common IO..................25
Figure 7 Demonstrator architecture..................28

Chapter 4
Figure 1 Analog (left)/digital (right) beamforming (transmitter)....32
Figure 2 DBF system and RF chip architecture (top), chip photograph (bottom)..................34
Figure 3 Receiver architecture from the antenna T/R interface to the BB VGAs..................36
Figure 4 BB VGA schematic..................37
Figure 5 SAR ADC block diagram.The proposed ADC..................38
Figure 6 Output spectrum for 25.1MHz input signal sampled at 250 msps..................39
Figure 7 Tx architecture..................39
Figure 8 PMU architecture..................40
Figure 9 Up-converter mixer architecture. 41
Figure 10 Power amplifier architecture and VGA first stage..................42
Figure 11 Tx DAC simplified schematic..................43
Figure 12 Tx DAC IM3 without DEM (red) and with DEM (blue)...44
Figure 13 Frequency multiplier block diagram..................45
Figure 14 Amplifier stage..................45
Figure 15 IO-structure..................47
Figure 16 Eye diagram for one of the I/Q channels receiving QAM16 @ 2.95 GS/s..................49
Figure 17 Power consumption (typical) distribution per RFIC in mW (top Tx mode, bottom Rx mode). Rx mode consumption is given in regular receiving mode, not in DPD mode..................50

Chapter 5

Figure 1 Maximum link data rate as a function of carrier frequency according to the Shannon capacity limit for a 20% fractional bandwidth channel. 58
Figure 2 Channel bonding TX architecture. 59
Figure 3 BB to IF band up-converter and channel bonding circuit implementation. 60
Figure 4 D-band TX and RX modules and circuits. 61
Figure 5 D-band channel bonding TX and RX module frequency domain characteristics. 61
Figure 6 LO signals spectrum and PN for the BB-IF up-converter. 62
Figure 7 LO signals spectrum and phase noise for the D-band TX and RX circuits. 62
Figure 8 Point-to-point links implemented using the D-band TX and RX modules described in Section 3.2. 63
Figure 9 D-band point-to-point link results. 64
Figure 10 BB to D-band TX block diagram. 64
Figure 11 BB to D-band point-to-point link set-up. 65
Figure 12 EVM versus channel RX signal power for the set-up shown in **Figure 1**1 as the TX-RX link distance is varied. 66
Figure 13 RX signal waveform, spectrum and constellations for a link distance of 35 cm and 56.32 Gb/s of total data-rate. 66

Chapter 6

Figure 1 System diagram of 39GHz transceiver frontend. 72
Figure 2 BEYOND5 packaging concept. (a) Cross section, (b) top view. 74
Figure 3 Possible setup concepts with the corresponding calculated temperature rise. 75
Figure 4 Packaging concept for standalone block RFIC. (a) Cross section, (b) top view. 76
Figure 5 Impact of etching tolerances on signal performance. 77
Figure 6 Impact of etching tolerances on signal performance. (a) Transmission parameters, (b) reflection parameters. 77

Figure 7 (a) Schematic of balling process of the interposer bottom side, (b) schematic of assembly process of the interposer top side. 79
Figure 8 Microscopic image of an underfilled MMIC. 80
Figure 9 (a) RF paths simulations, (b) manufactured interposer, (c) after reflow soldering, (d) soldered capacitor, (e) X-ray image for solder quality control. 81

Chapter 7
Figure 1 V2X ESPAR antenna model: (a) top view, (b) side view. 89
Figure 2 V2X ESPAR antenna DoA error estimation for three directional radiation patterns that can be rotated in the horizontal plane [16]. 89
Figure 3 Prototypes of antenna with lenses (antenna dimensions *D* × *H*): (a) extended hemispherical (70 × 86.1 mm), (b) ellipsoidal lens (70 × 91.0 mm), (c) cylindrical lens (70 × 88.1 mm), (d) GRIN lens (100 × 88.1 mm), (e) polarizer lens (73.4 × 101.1 mm). 90
Figure 4 Measured realised gain radiation patterns of antenna with different lenses: (a) E-plane measurement, (b) H-plane measurement. 90
Figure 5 (a) Antenna array model top view. (b) Antenna array with lens and spacer (antenna dimensions with lens: 120 × 120 × 82 mm). 91
Figure 6 Simulated 3 dB gain beams coverage at 5.9 GHz for all possible beams. 92
Figure 7 ERES antenna for 5.9 GHz. (a) Antenna model with one metallic wall removed. (b) Bottom side of the reconfigurable layer (antenna dimensions: 42 × 42 × 40 mm^3). 92
Figure 8 Simulated radiation patterns of the ERES antenna in the H-plane. 93
Figure 9 Simulated radiation patterns of the ERES antenna in the E-plane: (a) $\phi = 90°$, (b) $\phi = 0°$. 93
Figure 10 Fabricated antenna: (a) array top view, (b) array bottom view, (c) array with the lens mounted on a turntable in the anechoic chamber (antenna dimensions 55 × 41 × 5 mm, with lens 55 × 41 × 26 mm (W × L × H)). 93

Figure 11 Simulated half-power gain contours at 39 GHz for all possible beams....94
Figure 12 The antenna-in-package stack-up and an example of connections between the microstrip patches and transceiver modules....94
Figure 13 The design of the BGA interface (footprint of the array module)....95
Figure 14 The working principle of the antenna: (a) design of the reconfigurable microstrip to slotline transition, (b) surface current distribution in the R configuration – the R diode is forward biased....95
Figure 15 (a) Photograph of the manufactured antenna with soldered DC bias wires to control the switching state of the antenna. The unused solder pad is left to maintain the symmetry of the structure. (b) Comparison of simulated and measured normalized radiation patterns (dB) in the horizontal plane ($\theta = 90°$) for both switching states at the frequency equal to 60 GHz....96
Figure 16 Feed network steering points positions for the proposed antenna. In every considered antenna configuration, four points are shorted (marked red), while the remaining two are left open....96
Figure 17 Simulated 3D antenna radiation patterns....97
Figure 18 2D theta cuts of the simulated antenna radiation patterns....97

Chapter 8
Figure 1 Typical PA specifications for 5G and WiFi mobile applications....104
Figure 2 Ft vs. BV of state-of-the-art LDMOS devices in SOI CMOS technologies....105
Figure 3 State-of-the-art ETMs in CMOS technology: (a) ETM efficiency vs. bandwidth (bubble size: peak output power), (b) ETM efficiency characteristics vs. output power (extracted from [29], [30] and [33])....106
Figure 4 SOI LDMOS ETPA output stage: simulated PAE vs. OBO (Freq = 2.5 GHz, P_{sat} = 33 dBm)....106
Figure 5 SOI LDMOS DPA output stage: simulated PAE vs. OBO (Freq = 2.5 GHz, P_{sat} = 33 dBm)....106

Figure 6 Average PAE of DPA and ETPA for different modulation (DFTs-OFDM and CP-OFDM), cresting factor (0.01% and 1%) and shaping functions (efficient and linear).....107
Figure 7 Schematic and design equations of a Class-E SOI DPA with compact L-C combiner [44].....107
Figure 8 NR-ACLR (left) and PAE (right) vs. P_{out} for a 5G-NR 100 MHz full-RB QPSK signal (8dB PAPR) at 2.3 GHz and V_{dd} = 5 V.....108
Figure 9 EVM vs. output power at 2.4 GHz and V_{dd} = 3.85 V for a MCS13 BE40 WiFi signal.....109

Chapter 9
Figure 1 System diagram of the V2X FEM SoC.....117
Figure 2 Illustration of the input stage of the observation receiver.....118
Figure 3 Left: filter gain response for 120 MHz f_c, right: filter frequency response for 10 dB gain.....119
Figure 4 Pipeline ADC.....119
Figure 5 Spectrum of the ADC output.....120
Figure 6 Digital pre-distortion sub-system.....121
Figure 7 AC-coupled hybrid supply modulator.....121
Figure 8 Supply modulator (left) efficiency and (right) waveforms.....122
Figure 9 Left: layout of the example SoC, right: EVM simulation results.....122
Figure 10 The spectrum of the signal at the PA output.....122

Chapter 10
Figure 1 Depiction of the different substrate options (in the bulk and at the interface) and the overlying BEOL metal stack.....129
Figure 2 Effective resistivity (ρ_{eff}) and line losses (α) from the measured CPW lines (QB metal layer) on the different considered substrate options.....131
Figure 3 Inductance (L) and normalized quality factor (Q) of 5–20 GHz inductors on different substrates, with or without interface passivation.....132
Figure 4 Mm-wave SPDT switch design.....133

Figure 5 Microphotograph of the fabricated SPDT module. The measurement reference planes after open-de-embedding are shown as dashed lines. 133
Figure 6 Measured and simulated S_{21} parameter of the SPDT (open-de-embedded) as a function of frequency on all substrates in both the on- and off-states. 133
Figure 7 Measured and simulated SPDT insertion loss (open-de-embedded) on the different substrate options. 134
Figure 8 Schematic of the three-stage LNA design and microphotograph of the fabricated circuit. 135
Figure 9 Measured and simulated performance – noise-figure and gain – of the designed LNA fabricated on the different substrate options. 136
Figure 10 Schematic of PA1 (a), the driver stage (b) and power stage (c). 137
Figure 11 Microphotograph of the fabricated standalone output balun. 138
Figure 12 (a) Simulations (green) and measurements (black) of the standalone output balun on Std-POR. (b) Simulations of the output matching network on Std-POR. 138
Figure 13 Microphotograph of the full two-stage PA design. 138
Figure 14 Measured performance – *Pout* and *PAE* – of the designed PA fabricated on the different substrate options. 139

Chapter 11
Figure 1 Block diagram of the proposed human identification algorithm using MIMO FMCW radar. 145
Figure 2 Data acquisition setup with two human subjects. The radar and corner cubes are both placed on tripods. 147

Figure 3 Empirical results of the test scenario involving two humans: (a) the range-profile derived from one of the receiving antennas, marking the positions of the test participants and corner cubes; (b) the unprocessed RAI with indistinct coordinates of the human subjects; (c) the unprocessed time-evolving range-profile matrix; (d) the normalized amplitude of 1D curve-length metric; (e) the processed RAI in which the first and second participants are detected at (2.1(m), 20°) and (2.8(m), –35°), respectively; and (f) the time-evolving range-profile matrix after employing the DC-value suppression technique............148

Figure 4 Trajectory plot of the normalized IQ-values over CPI (4*s*) after employing the DC-value suppression method. The magenta markers correspond to the test participant's coordinates, while the green point relates to one of the corner cubes utilized.............149

List of Tables

Chapter 3
Table 1 EVM distribution in the access point transmitter.............25

Chapter 4
Table 1 Recent mmwave beamforming rfics................................33
Table 2 Performance in Rx mode...37
Table 3 Block power breakdown..48
Table 4 Bitrates for supported bitrates and modulations...............49
Table 5 Estimated energy per bit for supported bitrates and modulations...49

Chapter 5
Table 1 Link budget analysis..65

Chapter 7
Table 1 Comparison of measured parameters of realised antenna with lenses...91
Table 2 Design summary...98

Chapter 8
Table 1 State-of-the-art 5G-FR1 PA. [5g signal: dfts-ofdm qpsk 100 mhz)]...109
Table 2 State-of-the-art 2.4 ghz wifi pa....................................109

Chapter 10
Table 1 Substrate impact on the main figures of merit of the studied mm-wave devices...................................139

Chapter 11
Table I Simulation parameters of the radar-based human identification algorithm.......147

RF Technology Roadmap for 5G and 6G RF Front-end Systems

Yvan Morandini

***Abstract*—The evolution of 5G and 6G wireless systems adds considerable challenges to new RF front-end designs and their implementation. For 5G FR1 (450 MHz to 6 GHz) and FR2 (20–60 GHz), there is a significant increase in bands with more band combinations and extensions of the spectrum at higher frequencies. Beyond 5G, new FR3 (7–24 GHz), FR4 bands (>90 GHz) allocations are calling for disruptive semiconductor solutions. In this paper, we will provide a roadmap for RF technology including the engineered substrates enabling solving design challenges of wireless communication systems.**

***Index Terms*—Engineered substrates, RF front-end, RF-SOI, FD-SOI, POI, GaN and InP.**

I. INTRODUCTION

THIS publication focuses on 5G and beyond generation cellular front-end (RFFE) design and how material and engineered substrate development addresses the new wireless system requirements. In the first part, it will introduce the 5G RFFE evolution and its challenges. In the second, it will present the value of engineered substrates through RFFE 5G frequency range (FR) design, and in the third, it will discuss how some of the experience gained with engineered substrates could be leveraged from 5G FR1 into 5G FR2/FR2+ from 20–70 GHz. Finally, we will introduce a technology roadmap to address new frequency bands beyond 5G.

A part of this work has been supported through Important Project of Common European Interest (IPCEI) Microelectronics, Nano 2022, under French financial support, and from KDT Joint Undertaking (JU) under grant agreement No 876124 – project BEYOND5.

Y. Morandini is with Soitec, Bernin, France (e-mail: yvan.morandini@soitec.com).

II. 5G RFFE EVOLUTIOIN AND CHALLENGES

The current 5G frequency spectrum is introduced in Figure 1, including the WiFi spectrum: 2.4, 5.8 and now 6 GHz. With the aim of supporting Gbps data rate, the cellular RF front-end architecture will be more complex, supporting Wifi/5G coexistence, new bands and dual connectivity support.

In the FR1 spectrum, 5G sub-6 GHz bands feed the data demand required that 4G bands could not serve (risk of network saturation, resulting in poor quality of service).

Three new sub-6 GHz bands above 3 GHz and others are progressively moved from 4G to 5G (we are talking about band refarming). These new bands double the bandwidth compared to 4G. They will require additional receive and transmit modules as illustrated in Figure 2. In FR2, 5G millimeter waves (mmWave) bands feed the data demand that 5G sub-6 GHz bands alone do not serve. First mmWave bands are in the 28 and 39 GHz frequency range adding 10 more bandwidth than the 4G

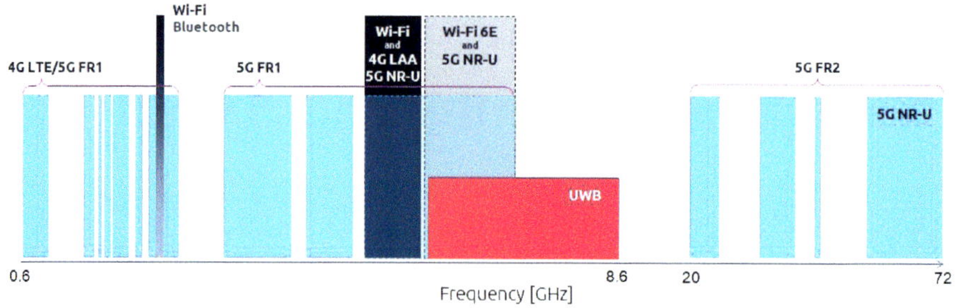

Fig. 1. 5G and WiFi spectrum.

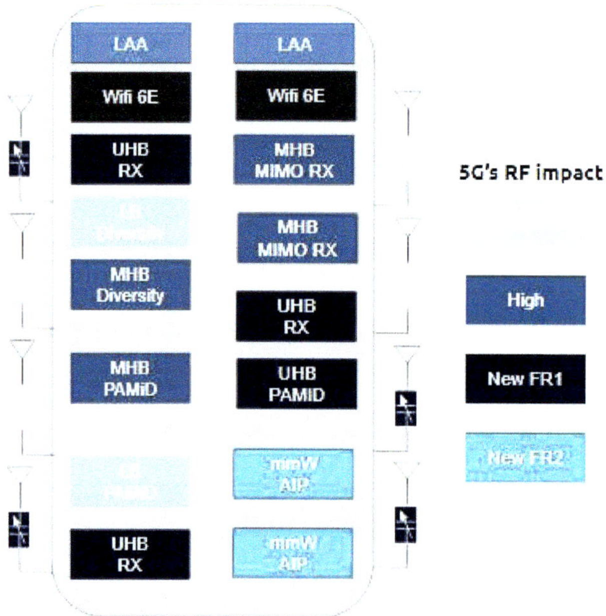

Fig. 2. RFFE module content for a 5G mobile handset.

bands. Additional bands will come in the future with 52 GHz and potentially 71 GHz in the long term horizon. Given that mmWave is a new paradigm, it will require a dedicated RF front-end with specificities in terms of architecture. Due to the smaller form factor of the antenna and with the aim of limiting the losses between antenna and RF front-end, which are much higher at mmWave frequency, the antenna and the RFFE could be integrated in the module. This is called an antenna in package (AiP). Comprehensive study of the user hand effect on the design of 5G mmWave mobile handsets specifically in terms of antenna module design and placement has been demonstrated. For that reason, depending on the phone design, two to three modules will be distributed over different locations within the mobile device.

Another 5G RFFE evolution is the fact the 5G mobile architecture needs to support dual connectivity, mainly for the first phase of the 5G deployment without a 5G standalone network and still with a LTE network anchor. Dual connectivity is a (non-standalone) feature that allows mobile devices to utilize both 5G and LTE bands to provide improved coverage and data rates. Release 15 of 3GPP has standardized the non-standalone feature in the aim to support connectivity to 4G and 5G networks, which is a requirement.

Finally, 5G smart phones need to support 4 × 4 multiple input multiple output (MIMO) architectures for the downlink and 2 × 2 MIMO for the uplink. Multi-antenna technology has become part of modern wireless systems. A 4 × 4 MIMO device means that the device presents four antennas to generate four data streams. Each antenna enables receiving and sending data. The increase in the number of antennas allows transferring more data with faster downlink and uplink speeds. MIMO 4 × 4 cellular architectures are widely deployed for 4G LTE advanced high-end mobiles, and will be mandatory for all 5G mobile devices. The 4 × 4 MIMO pervasion and the introduction of new 5G bands both FR1 and FR2 will introduce more RF front-end complexity in terms of number of 5G RF modules and 4G/5G coexistence.

The 5G RFFE evolution and transition has some challenges [1]:

- To achieve coexistence with multiple WiFi, UWB and Bluetooth, it is necessary to increase the antenna count to 6–8. These antennas should be accessible from various 4G/5G LTE radios with intra-coexistence with 4G bands in 5G bands.
- Support GSM and 4G/5G data requiring more linearity through multiple transmitter and receiver paths operating at the same time.
- New 5G dedicated bands and more carrier aggregation will require more filter paths putting more strain on the RF bill of material footprint (Figure 3).
- Bandwidth up to 100 MHz regarding FR1.
- 5G higher power class devices require up to 26 dBm at the antenna. This new power class category has been defined through 3GPP. It defines the maximum power of the transmit part over the full 5G NR channel bandwidth. It solves the problem of the limiting factor of the link budget: TX power from the user

FR1 CA combinations	Rel-16	Rel-17	
Intraband	19	30	
2 bands	61	176	
3 bands	35	182	
4 bands	6	75	
5 bands	-	2	
With SUL	15	52	
Total	156	519	> 3X

FR1 EN-DC combinations	Rel-16	Rel-17	
Intraband	27	32	
2 bands	371	517	
3 bands	845	1627	
4 bands	629	1465	
5 bands	184	484	
6 bands	7	36	
Total	2063	4161	> 2X

LTE CA combinations	Rel-16	Rel-17	
Intraband	39	39	
2 bands	260	290	
3 bands	292	336	
4 bands	112	163	
5 bands	9	33	
6 bands	-	3	
Total	732	864	+132

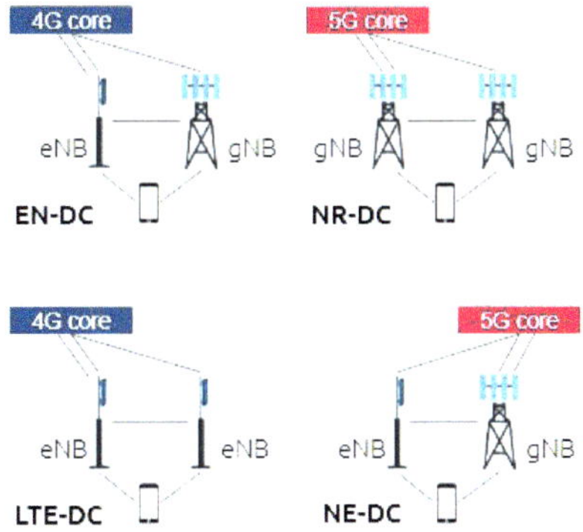

Fig. 3. 4G and 5G carrier aggregations.

equipment to the base station. When user equipment transmit power is high it enables better cell coverage. This means higher power amplifiers without efficiency degradation.

- Improve the power amplifier efficiency below 10% at 9.6 dB back-off for mmWave FR2 radios.

III. ENGINEERED SUBSTRATES FOR 5G FR1

In 2020, IMT set the vision and requirements for 5G technology defining the data rate to achieve from 100 Mbps on downlink down to 50 Mbps on uplink. Moreover a minimum network traffic capacity per area has been defined around 10 Mbps/m^2 [2]. The aim of achieving such data rates will require embedding from 80 to 100 MHz of the contiguous middle band from the 1–6 GHz spectrum per mobile network operator (MNO). A few component carriers (CC) of the contiguous and non-contiguous spectrum could be assigned to a user by a MNO in order to obtain, adding their individual bandwidths (BWs), a required channel BW and consequently a target data rate. With the move to 5G, the number of CCs and their potential combinations increases tremendously [3].

The higher RF complex signal modulation (higher order) and more carrier combinations have had major impact on the RFFE linearity specifications. It grows the importance of addressing RFFE filtering and linearity at an early stage of the design. Figure 4 illustrates the risk of interference between the N12 and N66 bands and the de-sensing of the low noise amplifier (LNA) when there is not sufficient substrate isolation. Linearity requirements such as lowering H2 and H3 (second and third harmonics) are better addressed from the initial phase of the RFFE design by choosing the right substrate technology for minimizing the interference. Figure 5 shows that RF-SOI substrates benefit in terms of harmonic performance, particularly those with a trap-rich layer, helping minimizing the third harmonic. Numbers as low as –90 dBm on a coplanar

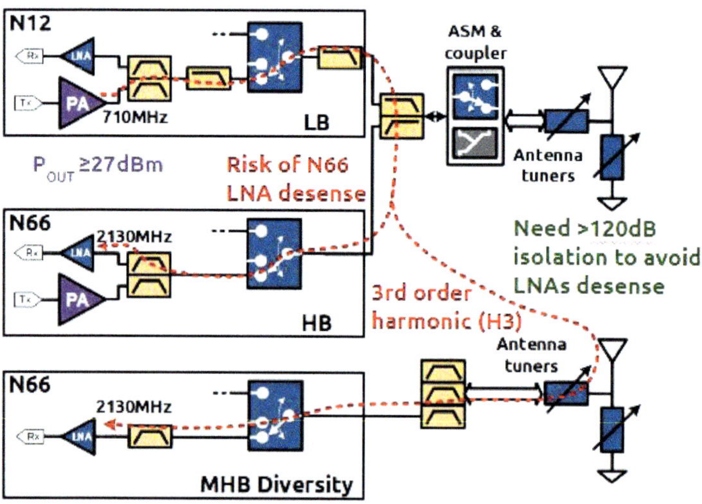

Fig. 4. RFFE risk interference N12 710 MHz and N66 2130 MHz bands example.

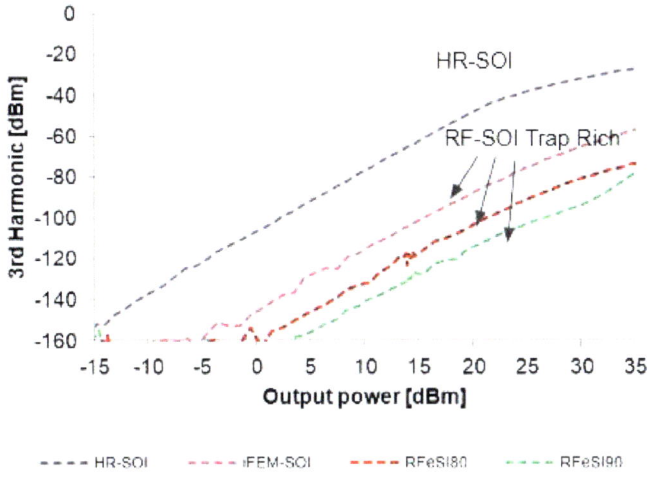

Fig. 5. Third harmonic as a function of the output power comparing HR-SOI with three trap-rich RF SOI flavors with resistivity range from 1 kOhm.cm to 20 kOhm cm at 900 MHz.

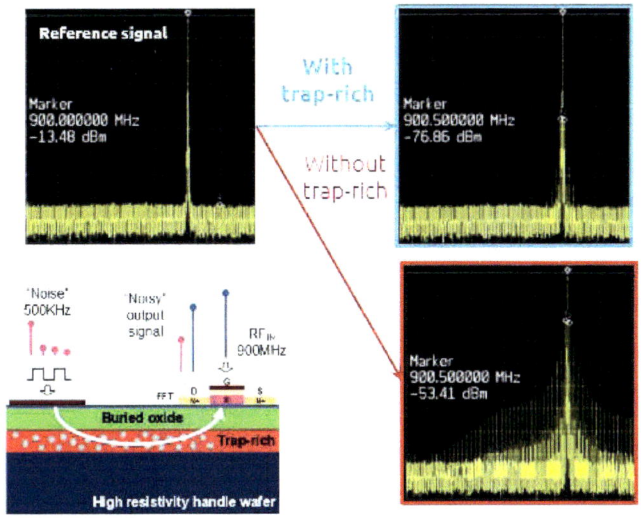

Fig. 6. Trap-rich RF-SOI digital noise and comparison with HR-SOI without trap-rich.

waveguide at 30 dBm output power P_{out} can be obtained.

Increasing the logic content is required to control the operation of integration of RFFE. This is true for 5G, Wi-Fi and other connectivity. It increases the challenges related to the optimization of the RFFE footprint and crosstalk isolation in a compact form factor (e.g., mobile). Figure 6 shows how RF-SOI substrates help with crosstalk and digital noise isolation while enabling optimal functional integration per RF die [4]. The test structure consists of a metallic pads located at a fixed distance from RF SOI n-MOSFET (Figure 6). The 350 μm distance between the noise injection RF pads and the transistor was calculated based on the minimum distance, fixed by the design rules to separate two orthogonal RF pads, which is 200 μm for our three-port RF measurement setup. The on-wafer measurements were performed using a Cascade Probe station with an Agilent E8267D RF generator which provide a 900 MHz RF input signal at –13 dBm and Agilent E4440A spectrum analyzer connected to the drain output port. A clock noise frequency of 500 kHz (5V) is injected through the noise RF pads 350 μm far from the transistor. The recorded output spectrum is measured at the drain port of the transistor. Although it has been expected that the buried oxide is enabling high isolation at low frequencies, the presence of a parasitic conductive surface layer at the buried oxide and high resistivity substrate leads to coupling and propagation of the large noise signal. However, in the case of the TR-SOI wafer the directly coupled noise is highly reduced. The largest noise peak at 500 kHz is –53.41 dBm whereas it is reduced to –76.86 dBm on TR-SOI.

Adoption of massive input massive output in receiver and transmitter front-ends, results in a higher numbers of bands, more carrier combinations and ultimately signal paths translate into a significant increase in the number of required RF filtering elements in the RFFE. Piezo-on-insulator (POI) substrates present a promising solution to accommodate the growing content within the same or smaller footprint. They offer the advantages of maintaining stable high rejection and low loss even at high temperatures. Moreover, they reduce the complexity of the manufacturing process [5]. The next figure illustrates the value of these POI substrates. As Figure 7 shows, POI allows state-of-the-art performance as compared to alternative technologies and POI has the capability to cover a larger spectrum than other competitive technologies.

In summary, POI enables the design of filters with high quality factor, large bandwidth, very low temperature sensitivity and low insertion.

IV. ENGINEERED SUBSTRATES FOR 5G FR2

More studies [6] highlight that an additional mmWave spectrum has been required in the aim to achieve the required cellular network capacity. The mid-band

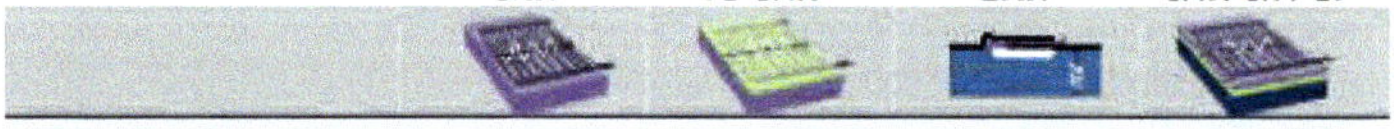

	SAW	TC-SAW	BAW	SAW on POI
Operating Frequency (GHz)	Low - High	Low – High	High Ultra High	**Low - High Ultra High**
Quality Factor (Bode Q) @2GHz	1500	<1500	4000 (AlN)	**>4000**
Coupling factor k^2	<7% ($LiTaO_3$)	<7%-$LiTaO_3$ <20% -$LiNbO_3$	<7% - AlN <12% - AlNSc	>8% - $LiTaO_3$ >20% -$LiNbO_3$
Temperature Compensation (ppm.K^{-1})	>40 - $LiTaO_3$	<20 - $LiTaO_3$	>20	< 10 - $LiTaO_3$
Substrate complexity	Low	Low	Low	**High**
Process complexity	Low	Mid	High	Low
Integration	Low	Low	Low	High

Fig. 7. Piezo-on-insulator (POI) versus alternative RF filter technologies benchmarking.

spectrum would not be long enough in the future. In the FR2 spectrum, the RFFE limited range is partially overcome by introducing spatial power combining techniques. Depending on final applications – mobile user equipment (UE), customer premises equipment or small cells – the same range in terms of coverage (both transmit and receive) is not needed. An array that forms a group of elementary antennas is phased in the aim to concentrate the transmitted power through a narrow beam. This is called beamforming. Relative to transmitting the same total power from one antenna, the array achieves an added gain of 10logN where N is the number of elements in the array.

For mobile, the user equipment power level is smaller and will require between four and eight antennas. However, the form factor of the mobile user equipment and the user interaction with the environment strongly constrains the practical implementations. Key considerations include the aspect ratio of the UE platform chassis, the display coverage of one or both large area sides, the cover materials, and the collocation with the sub 6GHz antennas. Achieving the desired RF performance will necessitate finding a balance between power transmission and receiver sensitivity, while also considering the level of digital integration. The first architecture, Figure 7(a), is suitable with the form factor of the UE and is well established.

Some other equipment (Customer Precise Equipment) with different form factor and with higher power transmit requirement could benefit from less integration. More power to transmit will translate to larger phased antenna arrays (up to 64 antennas). The up/down converter will not be integrated offering more flexibility considering best in class RF technology for the RFFE (Figure 8(b)).

Given, the power requirement of the small cells and the less stringent form factor constraints associated with it, the small cells could benefit from the disintegration of transmit and receive part of the RFFE. Figure 8 shows the different options and level of integration of a mmWave RFFE. The architecture on the left is well-suited for striking a balance between RF performance and digital integration. The utilization of a commercial low-leakage FD-SOI RF planar CMOS platform has allowed the successful demonstration of this mmWave RFFE architecture in the mobile market versus CMOS bulk technology improving

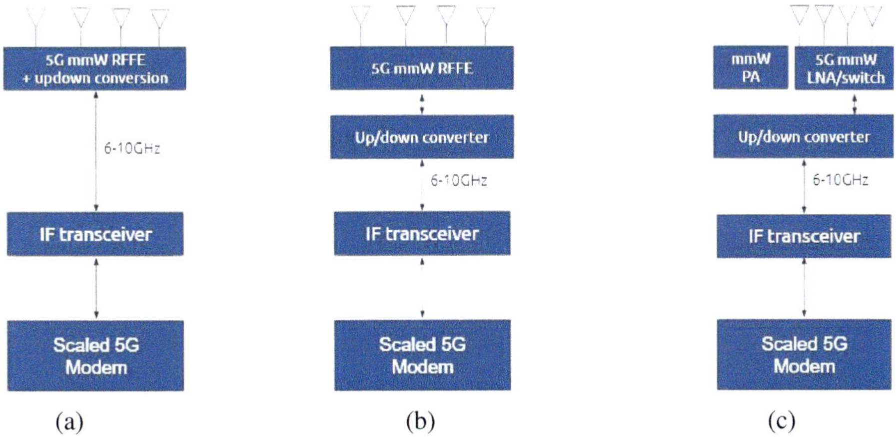

Fig. 8. mmWave RFFE integration options.

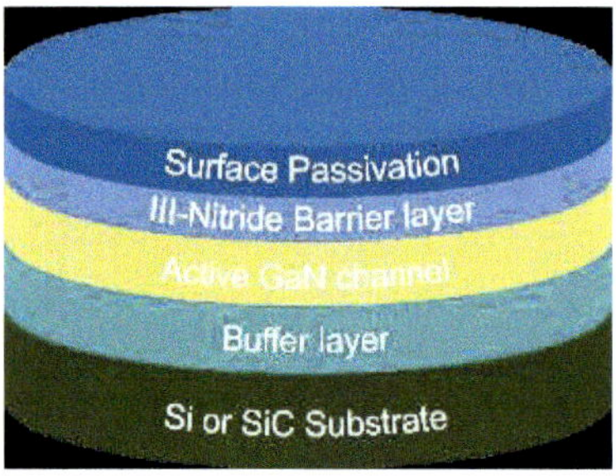

Fig. 9. Epitaxy GaN layer.

power amplifier efficiency, noise figure of low noise amplifier and optimized power consumption [7].

The middle architecture of Figure 8 has been implemented in commercial products for 5G customer premises equipment (CPE) and point-to-point radio communication applications [8]. To achieve array antenna scalability, the 5G mmWave front-end incorporates an eight-channel beamforming front-end along with separate dual-channel up-/down-converter ICs that are not integrated with the RFFE.

For external small cell equipment more suitable for less constraint integration but higher power and DC power constraints, III–V materials are an alternative to CMOS technologies with better saturation power, linearity and power efficiency for large back-off. The short gate length of GaN components [9] has demonstrated good performance for mmWave power amplifiers. With higher operating voltages and reduced device parasitics, these transistors provide higher output power densities, wider bandwidths and improved DC-to-RF efficiencies than their GaAs counterparts. mmWave GaN technology requires an optimized epitaxy stack.

The ultra-thin InAlN barrier layer shows a higher F_t and F_{max} at higher current densities [9]. This follows the enhancement of the transconductance at higher current densities. The HEMTs with an InAlN barrier outperform the AlGaN barrier devices at large signal, largely thanks to their higher I_{max}, which strongly impacts the maximum possible P_{out}.

Figure 10 with an InAlN barrier show a high gain of 19 dB while AlN shows excellent PAE of 71% and 5.2 W/mm output power in pulsed mode compatible with small cells.

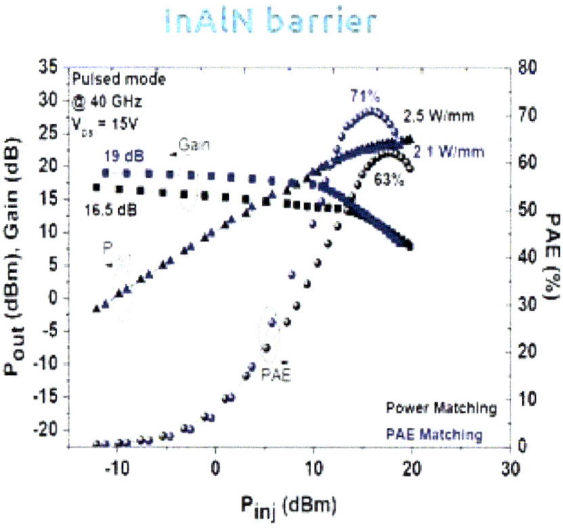

Fig. 10. Load-pull measurement of a HEMT device with InAlN epitaxy barrier.

V. ENGINEERED SUBSTRATES FOR NEXT PHASE OF 5G AND FUTURE 6G

Evolution of the 5G FR1 spectrum (extended below 7 GHz) and the transition to 6G FR3 (between 7 GHz and 24 GHz) requires technology improvement of the power amplifiers. In comparison with GaAs incumbent technology, RF-GaN on silicon technology offers the potential to optimize the RF footprint by facilitating the integration of higher power density transistors into mobile power amplifiers. Such transistors could also offer wide bandwidths, high output impedance and good linearity, bringing significant advantages to the mobile market [10].

In recent years, it has been demonstrated that the benefits of RF-SOI substrates could also be extended to CMOS alternatives such as GaN, providing them with the potential of being integrated with SOI CMOS technology offering state-of-the-art low loss/high isolation RF-SOI switches and high-quality factor, Q factor, interconnects, inductors/transformers and capacitors. GaN on RF-SOI provides an excellent alternative for optimizing the PA module footprint at mid-band and mmWave bands [11].

The sub-THz spectrum (FR4) beyond 100 GHz, including the D-band, has also gained quite some interest for achieving the ultra-high data rate goals of the next generation 6G cellular network. Carrier frequencies >100 GHz create significant challenges in term of power efficiencies. Semiconductor and compound technologies with transistor F_{max} values >500 GHz could be needed for improved efficiency, gain and noise. Even if SiGe technology has the potential to increase silicon transistor performance, InP technology has the best front-end performance at sub-THz with best power amplifier figure of merit P_{out} and power added efficiency (PAE) [12] and low noise figure [13] depending on the HEMT or HBT transistor. So far, in part due to the constraints related to its manufacturing with low diameter and expensive, it has had a limited adoption in markets with low volume production. For it to be successful in the consumer market, reliable and cost-efficient manufacturing should be secured. An innovative method to manufacture large diameter – up to 300 mm – cost-efficient InP wafers is described in [14] with wafer reconstructing and reclaiming.

VI. CONCLUSION

In this paper, we have demonstrated the importance of engineered substrates and their impact on the RFFE performance for next generation 5G and 6G applications. Substrate innovation and improved manufacturability present a crucial starting point for optimizing RFFE performance, offering potential benefits to the entire RF system. We have demonstrated the benefit of RF-SOI for switches in RFFE (extended to low noise amplifier and antenna tuner), FD-SOI for 5G mmW System on Chip, POI technology for 5G RF filters and GaN and InP for power amplifier solutions. All solutions has been demonstrated to be suitable for high volume manufacturing and are cost efficient for 5G and next 6G markets.

REFERENCES

[1] F. Balteanu et al., "5G RF front end module architectures for mobile", *EuMA* 2019.
[2] ITU-R, "Minimum requirements related to technical performance for IMT-2020 radio interface(s)," *ITU-R M.2410-0*, November 2017.
[3] J. Young, "SOI integration workshop," *IEEE S3S*, June 2014.
[4] L. Andia, Y. Morandini, J.M. Lemeil, "RF-SOI engineered substrates at the

heart of modern RF mmWaveave front-ends," *EE Times Europe*, November 2020, pp. 1-3.

[5] E. Butaud et al., "Smart Cut™ piezo on insulator (POI) substrates for high performances SAW components," *IEEE IUS*, 2020, pp. 1-4.

[6] J. Madden, "mmWaveave will be the critical 5G link," *Microwave Journal*, May 2019, pp 24-38.

[7] V. Bhagavatula et al., "A 5G FR2 power-amplifier with an integrated power-detector for closed-loop EIRP control," *IEEE Journal of Solid-State Circuits*, vol. 57, no. 5, pp. 1257-1266, May 2022.

[8] Psemi, "pSemi introduces complete 5G mmWaveave RFFE solution", *Microwave Journal*, June 2022.

[9] R. ElKashlanmm et al., "mm-Wave GaN-on-Si HEMTs with a PSAT of 3.9W/mm at 28GHz," *IMS*, June 2023

[10] Q. Yu et al., "A fully integrated 3.2-4.7 GHz Doherty power amplifier in 300 mm GaN-on-Si technology," *IEEE BCICTS*, 2022.

[11] B. Y. Wang et al., "Demonstration of patterned GaN RF MIS-HEMTs growing on hybrid oriented silicon-on-insulator (SOI) substrates," *IEEE DRC*, 2022, pp. 1-2.

[12] https://gems.ece.gatech.edu/PA_survey.html.

[13] Samsung, Workshop WSG IMS 2021.

[14] B. Ghyselen et al., "Large-diameter III–V on Si Substrates by the Smart Cut process: The 200 mm InP film on Si substrate example," *Phys. Status Solidi* A, Vol. 219, Issue 4, Feb. 2022.

Dr. Yvan Morandini received his Degree in engineering from Grenoble Alpes University, France in electronics and radiofrequencies in 2005 and his Ph.D. degree from Lille University in 2008. After 3 years in STMicroelectronics he then moved to IBM, Dolphin Integration and then to Soitec. He is currently Strategic Marketing Senior Manager at Soitec. He has 20 years of experience in the semiconductor industry including characterization, modeling and design. He has authored/co-authored over 15 papers in international, peer-reviewed journals and conferences, and is a Member of Institute of Electrical and Electronics Engineers (IEEE).

From Beyond 5G to the Sub-THz Era

François Brunier

***Abstract*—Today, the importance of semiconductor substrates for sovereignty and key applications competitiveness is rising. European leadership in key markets requires the establishment of a complete and reliable supply chain starting from the material. The "value chain model" in European programs such as the KDT "BEYOND5"* (building the fully European supply chain on RFSOI, enabling new RF domains for sensing, communication, 5G and beyond) is a motorway to accelerate the co-innovation pace and market adoption. This paper aims to illustrate the SOI (silicon on insulator) ecosystem innovation dynamic in RF (radio frequency) applications up to 120 GHz from the BEYOND5 project perspective. It proposes an opening on the sub-THz frequency range showing how the industry-proven Smart Cut™ technology can be adapted to other materials such as InP (indium phosphide) and applications.**

***Index Terms*—RFSOI, FDSOI, Smart Cut™, 5G, mm-wave, IoT, radar, THz, InP, value chain, pilot line.**

I. Introduction

In collaboration with CEA-Leti [1], Soitec [2] created more than 30 years ago the patented Smart Cut™ process allowing the manufacture of advanced substrates for microelectronics.

This process, which consists of transferring a thin active layer of silicon or another semiconductor material from a donor substrate to a second substrate acting as a support, as described in Figure 1, can be used in various electronic technology fields, such as RF components (RFSOI), low-power electronic components for computing and artificial intelligence (FD-SOI), photonics components (photonics SOI) or power components for automobiles (power SOI) and electric mobility (SmartSiC).

Using this technology, an unlimited source of innovation, as shown in Figure 2, gives Europe the ability to cover the entire value chain, and in particular the basic material, and thus limit dependencies. This Smart Cut™ technology has proven its value with the presence of RFSOI based components in all Smartphone front-end modules. Furthermore, it is noteworthy that the innovation and impact generated by this model has been recognized at a national and European level:

- December 2021: European EARTO Innovation Award for the innovative Smart Cut™ technology.
- December 2020: "Stars of Europe" award by the French ministry of higher education, research and innovation.

In the framework of the Important Project of Common European Interest (IPCEI) Microelectronics, the work takes place in Nano 2022 under French financial support. The project BEYOND5 "Building the fully European supply chain on RF-SOI, enabling new RF domains for sensing, communication, 5G and beyond" has received funding from KDT Joint Undertaking (JU) under grant agreement No 876124.

F. Brunier is with Soitec, Bernin, France (e-mail: francois.brunier@soitec.com).

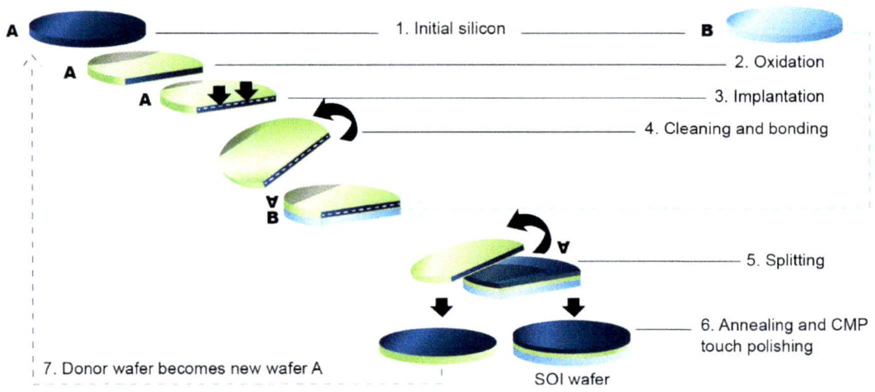

Fig. 1. Schematic illustration of the Smart Cut™ technology used to manufacture SOI substrates.

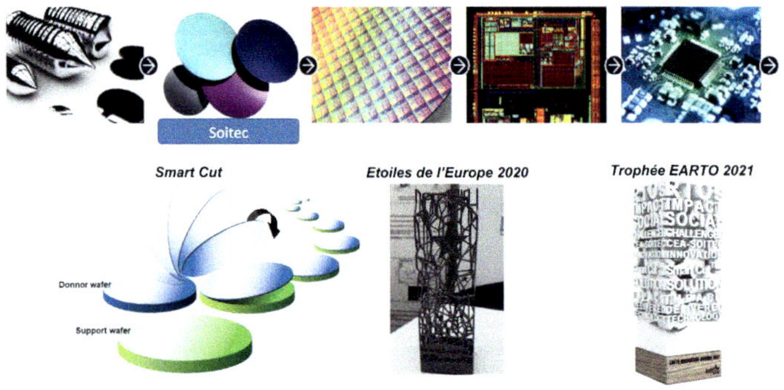

Fig. 2. Smart Cut™ technology, a source of innovation for a reliable, sustainable, less dependent value chain.

II. RF TECHNOLOGIES ON SOI, AN UNRIVALED PPAC FOR 5G AND BEYOND

RFSOI and FDSOI technologies based on SOI substrates not only meet the requirements of the global 5G smartphone market, but also offer new solutions for applications using the 5G frequency spectrum that stretches from low gigahertz bands all the way up to mm-wave deployments, including future prospective beyond 100 GHz.

With greater integration potential than alternative technologies, combining RF and digital for machine learning capabilities at the edge, high frequency performance and low power consumption, RFSOI and FDSOI are an opportunity to boost Europe's (EU) competitiveness at the forefront of 5G and beyond. The ability to integrate logic and analogy building blocks with FDSOI and RFSOI technologies brings a particular improvement in terms of PPAC for direct customers such as foundries, fabless and IDM (integrated device manufacturer).

The properties of RFSOI substrates, especially those with a trap-rich layer, help minimize third harmonic distortion, crosstalk and digital noise while enabling optimal functional integration per RF die [3]. These advanced substrates bring a competitive advantage and are present in 100% of smartphones with growing content at each generation.

Fig. 3. FD-SOI, an integrated platform selected by worldwide key players.

The European FDSOI technology is an integrated and versatile platform selected by key players in the communication, automotive and smart objects sector, as depicted in Figure 3.

Given the small device dimensions, FDSOI enables very high cut-off frequencies and therefore circuit operations at very high frequencies, in the mm-wave range and up to >100 GHz. Mm-wave RF front-end architecture using a commercial low-leakage FDSOI RF planar CMOS platform has been demonstrated in the mobile market [4].

In addition, FDSOI enables co-integration of analog and digital capabilities into system-on-chip and offers a dynamic tuning mechanism of the power/performance ratio thanks to the back-biasing capability which makes it key for automotive radar and automatic driving assistance systems (ADAS).

III. BEYOND5, A KDT VALUE CHAIN ON SOI FOR RF

BEYOND5, "building the fully European supply chain on RFSOI, enabling New RF Domains for sensing, Communication, 5G and beyond", generates an ecosystem dynamic on the RFSOI and FDSOI technologies. Figure 4 gives a visual description of this ecosystem by showing the different partners per do-

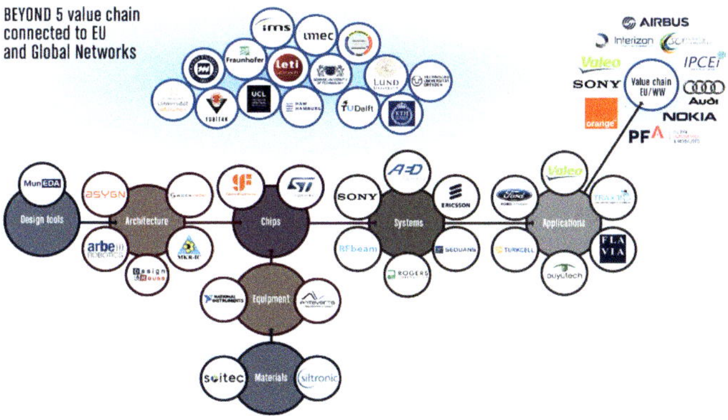

Fig. 4. KDT project BEYOND5 consortium.

main as well as the associated research institutes.

Started in June 2020, BEYOND5 gathers 39 European partners from 10 countries. The project aims to accomplish sustainable RFSOI and FDSOI RF platforms to cover the frequency range from 0.7 GHz to 120 GHz and prove the technical advantage of SOI in NB-IoT (narrow-band internet of things), V2X (vehicle to everything communication), 5G infrastructure, contactless USB and automotive radars (interior/exterior) demonstrators, as shown in Figure 5.

The project covers the whole value chain with three technological pilot lines lead by Soitec for the substrate, STMicroelectronics (ST) for the RFSOI technology [6] and GlobalFoundries (GF) for the 22FDX [7], and seven demonstrators driven by Ford, Traxens, ST, Ericsson, Valeo and Silicon Radar, aiming to showcase these technological platforms at the system level.

The project's substrate-level innovation includes development and integration of new trap rich layers for the RFSOI substrates and high resistive base wafer for the FDSOI substrates to further improve power amplifier efficiency and reduce the system level consumption.

In the RFSOI technology pilot line, advanced RF front-end (RFFE) modules for sub 6 GHz applications such as IoT and V2X are developed also as building blocks for 28 GHz compatible 5G RF front-end modules to prove the potential of this technology in the mm-wave frequency range.

In the FDSOI pilot line GF develops and integrates new RF functions and reliability methodology onto its 22FDX baseline, including the usage of high resistivity base wafers to improve 5G demonstrators' properties and future mm-wave systems requirements.

Each of the seven demonstrators chose the most competitive SOI technology based on its specific requirements. The main criteria are: frequency range, performance, integration, power efficiency and cost.

- For the 0.7 GHz IoT/asset tracking node, Traxens and IMS selected the RFSOI 130 nm technology for power efficiency, better robustness and cost.

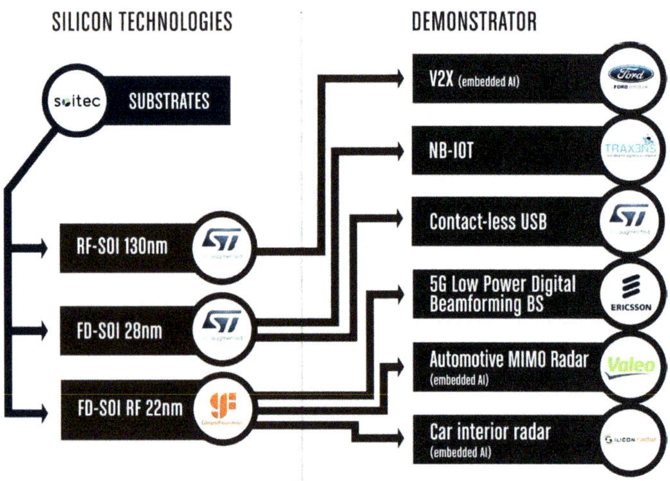

Fig. 5. KDT project BEYOND5 value chain.

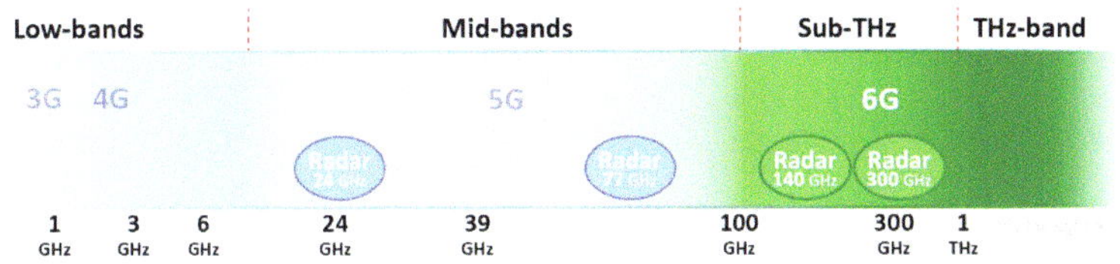

Fig. 6. Radio frequency range trend.

- For the 4.9 GHz V2X communication targeted to be deployed in an autonomous construction site by Ford, CEA-Leti and Tubitak selected an RFSOI 130 nm technology for power efficiency, better robustness and cost.
- For the 60 GHz mm-wave short range connectivity/contactless USB, ST selected the 28 nm FDSOI [8] which gives a competitive advantage in integration, performance and power consumption at a competitive cost.
- For the 39 GHz 5G low power base station, Ericsson [9] selected the 22FDX technology for its excellent co-integration capabilities of analog and digital building blocks, enabling architectural trade-offs to find cost viable solutions.
- For the 80 GHz automotive MIMO radar, Valeo and Arbe-Robotics selected the 22FDX technology for its high-speed RF transistors with high gain/low parasitic capacitance allowing high sensor performance with low-power and cost-efficient design.
- For the 125 GHz in-cabin radar/gesture recognition, Silicon Radar selected the 22FDX technology for RF performance allowing low-power designs with operating frequencies beyond 100 GHz and scaling effects allowing true low-cost designs for processing units (smallest area consumption) for SoP (system on package) solutions.

IV. SMART CUT™ APPLIED TO INP: A PATH TO THz

The Smart Cut™ technology brick is adaptable to other materials than silicon. By limiting the use of rare and expensive bulk material, it provides solutions for technological sovereignty compatible with Europe's Green Deal ambitions [10].

The electronics industry is pushing operating frequencies towards the THz range for even greater performance, throughput and reduced energy consumption. The future generation of mobile communication, 6G, is scheduled to use these frequency bands from 2030, as shown in Figure 6.

Indium phosphide (InP) has outstanding and unique capabilities to surpass other technologies in terms of performance and power consumption. The challenge is that indium is a rare and very expensive material whose mining reserves are 75% owned by China. The

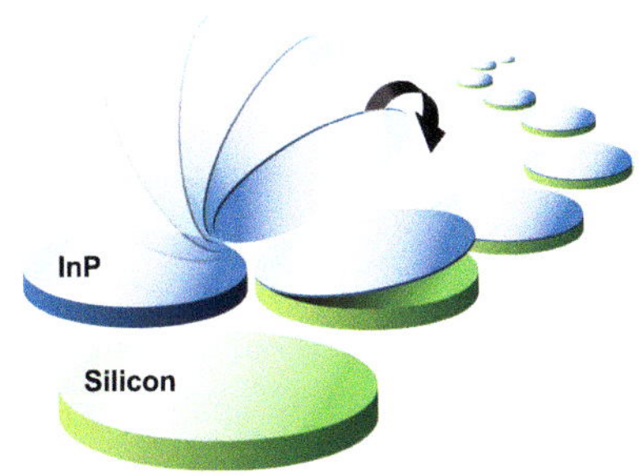

Fig. 7. Smart Cut™ of InP.

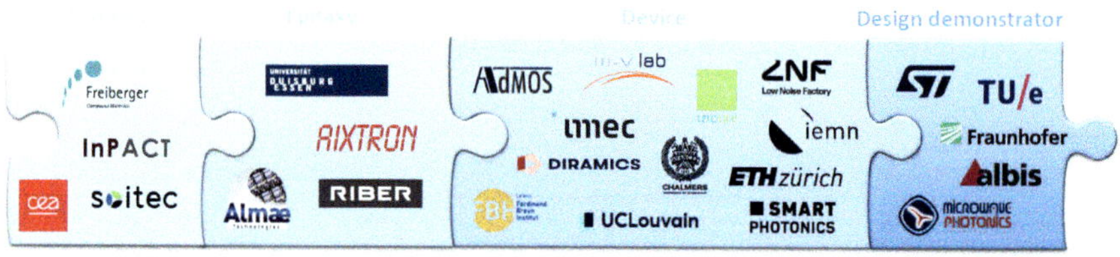

Fig. 8. Move2THz value chain.

Smart Cut™ model provides a sustainable and commercially viable solution to consumer applications, such as 6G mobile communication and sensors (image sensors, SWIR, biosensors, etc.). A smart InP technology platform, as depicted in Figure 7, would enable an increase in the number of components from the same InP wafer by a factor of 10 and to scale to larger wafer diameters using a tilling approach.

A new project proposal, "Move2THz", whose value chain is summarized in Figure 8, has been submitted to Call KDT RIA 2023. It brings together the European ecosystem on InP and will radically innovate the manufacturing process by establishing a breakthrough InP on silicon (InPoSi) global standard. This facilitates upscaling the wafer size and volume compatible with CMOS manufacturing capacities, while minimizing the use of rare InP resources. By establishing a complete and reliable value chain from the material, and by demonstrating its differentiation, Move2THz aims to transform the InP niche technology into a sustainable and commercially viable platform, enabling mass-market applications like 6G communication, photonics datacom, and RF-/bio-sensing to utilize frequencies towards THz and beyond.

V. CONCLUSION

European leadership in key markets requires the establishment of a complete and reliable supply chain that starts with the material. In this paper, we build the industry proven Smart Cut™ for the semiconductor industry, especially for low digital power consumption, sensing and telecommunications. We show that this Smart Cut approach is a kind of "toolbox" that can be adapted to different materials and open new application doors. By promoting the value chain model, KDT projects stimulate synergies and federate numerous players to create a veritable ecosystem along the supply chain. Consequently, this fosters the condition to accelerate the pace and adoption of new technologies to be time-to-market.

References

[1] https://www.leti-cea.com/
[2] www.soitec.com
[3] L. Andia, Y. Morandini, J.M. Lemcil , "RF-SOI engineered substrates at the heart of modern RF mmWaveave front-ends," *EE Times Europe*, November 2020, pp. 1-3.
[4] V. Bhagavatula et al., "A 5G FR2 power-amplifier with an integrated power-detector for closed-loop EIRP control," *IEEE Journal of Solid-State Circuits*, vol. 57, no. 5, pp. 1257-1266, May 2022.
[5] B. Ghyselen et al., "Large-diameter III–V on Si substrates by the Smart Cut process: The 200 mm InP film on Si substrate example," *Phys. Status Solidi A*, vol. 219, no. 4, Feb. 2022.
[6] S. Dhar et al., "Performance trade-off of RFSOI switches under scaled bias conditions," *2023 IEEE 23rd Topical Meeting on Silicon Monolithic Integrated Circuits in RF Systems, Las Vegas, NV, USA*, 2023, pp. 38-40

[7] M. Wiatr and S. Kolodinski, "22FDX™ technology and add-on-functionalities," *ESSDERC 2019 - 49th European Solid-State Device Research Conference (ESSDERC), Cracow, Poland,* 2019, pp. 70-73
[8] P. Magarshack, "Breakthrough technologies and reference designs for new IoT applications," *2015 Symposium on VLSI Circuits (VLSI Circuits), Kyoto, Japan,* 2015, pp. C42-C43
[9] https://www.ericsson.com/en/blog/2015/3/massive-beamforming-in-5g-radio-access
[10] https://commission.europa.eu/strategy-and-policy/priorities-2019-2024/european-green-deal_en

François Brunier graduated as a physics and electronics engineer from Centrale-Supelec in 1997. From 1998 to 2002, he worked as a device integration engineer for embedded DRAM products in STMicroelectronics Crolles. In 2002, he joined Soitec as head of Advanced Characterization Laboratory. From 2009 to 2011, as a product manager, he led the RF-SOI and power SOI product development and offering. Since 2012, as a partnership program manager, he has been in charge of European collaborative KDT programs, IPCEI and public relations.

A Low Power 5G Access Point Targeting Airplane Cabin Connectivity

Fredrik Tillman, Pallavi Paliwal, and Daniel Eckerbert

Abstract—**Today we see in practice how the fifth generation of mobile networks is transforming our society at global scale. End consumers can enjoy enhanced terrestrial mobile broadband services, including AR/VR experiences, and at the same time industries and factory floors use connectivity and cloud services to enhance efficiency and value creation. With the ambition to provide data sharing anywhere at any time, for anyone and anything, new use cases have emerged. In this paper we discuss opportunities for inflight 5G communication, and how a possible in-cabin 5G MIMO radio could be specified and implemented. A demonstrator in a real environment is further discussed, as well as constraints affecting the building practice and overall deployment.**

Index Terms—**5G deployment, specification, transceivers, low power, integration, 3GPP NR standard.**

The research leading to these results has received funding from the European Union's ECSEL Joint Undertaking under grant agreement n° 876124 – project BEYOND5. Numerous BEYOND5 consortium partners have contributed to the system realization and demonstrator planning presented here. A special thanks goes to involved partners in the demonstrator definition and implementation. These include Gdansk University of Technology, CEA-Leti, Fraunhofer IZM & EMFT, imec, MKR-IC, Asygn, TU Delft, NI, TU Dresden, Lund University, and KTH. We are also grateful to Airbus for fruitful discussions.

F. Tillman, P. Paliwal, and D. Eckerbert are with Ericsson Research, Ericsson AB, Lund, Sweden (e-mail: {fredrik.tillman, pallavi.paliwal, daniel.eckerbert}@ericsson.com).

I. INTRODUCTION

When observing the immense penetration of telecommunication in society, across all levels, it is of no surprise that the total mobile subscription count continues to grow. Despite geopolitical uncertainties, there are 228 commercial 5G networks active today and 35 already show public 5G SA (standalone) capabilities [1]. By 2028 it is envisaged that 5G will have surpassed 4G and break the 5 billion mark [2]. This is a five-fold increase compared to the beginning of 2023, meaning that more than half of all mobile subscriptions globally will be 5G at that time, as indicated in Figure 1.

In addition, fixed wireless access services (FWA) as a mean to enable cost efficient last mile network roll-out will continue to grow (see Figure 2) and reach 300 million connections by 2028. Among these, 5G will power more than 80%. It is also interesting to notice that the uptake of 5G is faster than the previous generations. In principle, 5G will reach the same penetration as 4G two years faster counting since first deployment. 4G on the other hand was faster than 3G, etc., which in turn indicates the growing importance of using the latest communication standard among consumers and business verticals. At the same time, the uptake for mobile PC and tablet subscriptions shows moderate growth, reaching around 680 million by 2028. One

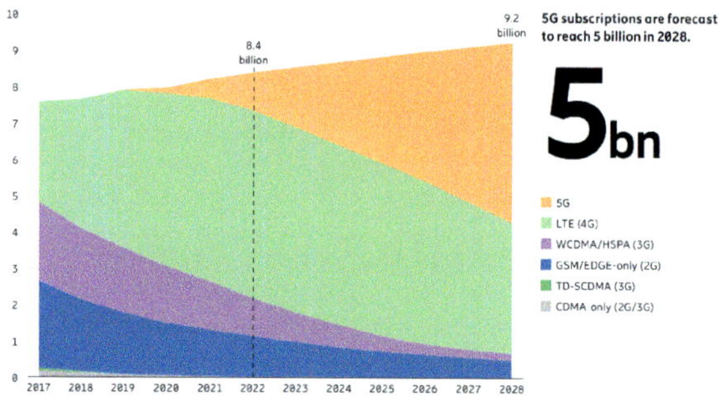

Fig. 1. Mobile subscriptions by technology (billion).

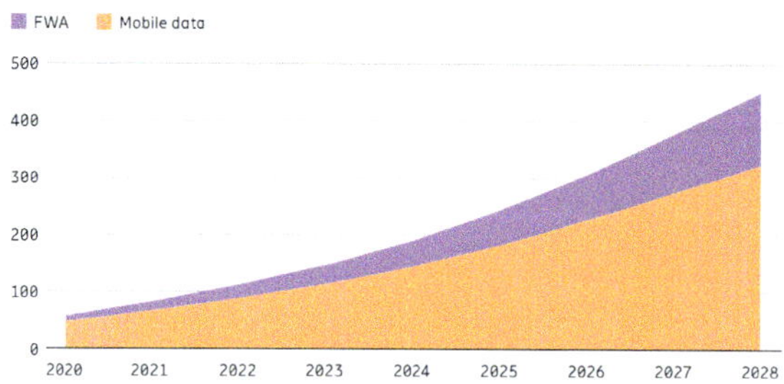

Fig. 2. Global mobile network data traffic (exabyte per month).

reason being today's user-friendly tethering and the massive growth of 5G enabled smartphones. The evolution of Wi-Fi has also enabled capable local networks where mobility is not in focus.

Zooming in on the access technology, a key difference between 4G and 5G is the introduction of mmWave frequency bands to capture more signal bandwidth. It ranges today from 24.25 GHz to 43.5 GHz in commercial deployments and is anticipated to expand further towards sub-THz frequencies in the future. The frequency step to mmWave has triggered massive R&D efforts for multiple years across the telecom community and resulted in new transceiver designs, beamforming techniques, channel propagation insights, etc. [3] [4]. However, despite all efforts to increase the downlink effective isotropic radiated power (EIRP) and employ larger arrays to handle uplink capacity, the market for mmWave connectivity is still dominated by FWA services and localized dense environments, including enterprises with ultra-reliable and low-latency communication (URLLC) needs [5] [6]. This is a natural business development as user coverage is challenging at mmWave frequencies given the excess pathloss compared to 4G bands. As a result, the investment costs per service area is higher, thus requiring enough user density to become business viable.

One example of a dense user space is public transportation. It represents a confined area with many users in capacity to consume data at high rates, e.g., streaming high-definition visual content or performing video interaction. In such a controlled and semi-static environment, with short distance communication, a mmWave deployment is logical as the benefit in terms of available bandwidth clearly outperforms the problems with coverage and mobility. In fact, some of the known drawbacks can be favorable and create opportunities, e.g., less interference between neighboring service areas sharing frequency resources and easy overlay for operators as mmWave is not an option for the macro deployment grid. Instead, challenges are shifted toward the use of massive MIMO, cost efficient computation, and hardware building practice with highly integrated antenna solutions. To investigate how a power efficient mmWave system operating at 39 GHz (NR260 band) could be designed for short distance reach, and to understand hardware realization challenges for deployed access points, a European project (BEYOND5) was established as part of the ECSEL-2019 framework [7]. The scope spanned across hardware design and system integration with the goal to provide a real environment over-the-air demonstrator. Today the activity has concluded needed block level designs, e.g., [8] [9], and currently targets the final demonstrator integration. The remainder of this chapter will discuss and provide a background to this work and associated specification aspects.

II. USE CASE AND SYSTEM REQUIREMENTS

The selected use case for the demonstrator was an airplane cabin environment as it contains several important aspects. Firstly, it's a confined user dense environment with low user mobility and a short distance between the access point and the connected equipment. This makes the radio environment suitable for mmWave operations. Secondly, general constraints of aviation enforce strict requirements on building practice (size, weight, power supply) and heat dissipation. As a result, the optimization point will be different compared to factory floors, FWA installations or public hotspots. The third aspect is co-existence with other networks. As airplanes are moving objects, the cabin network will be exposed to other network deployments unconditionally, e.g., cellular services covering an airport tarmac. This means in practice that large interferers may be generated by equipment located close to the in-cabin access point when connected to a terrestrial network. All in all, the in-cabin connectivity use case offers many relevant challenges for the system that are also applicable for other scenarios.

Deployment Considerations

To set the capacity ambition in 2019, the NGMN Alliance 5G white paper from 2015 served as a reference point [10]. It assumes that large aircraft (400 passengers) will need an aggregated downlink/uplink speed of 1200/600 Mbps. With an anticipated user activity factor of 20% this results in a 15/7.5 Mbps experienced user data rate. In the 3GPP rel-16 NR standard, the minimum radio bandwidth is 50 MHz for mmWave bands (FR2). Applying a 64 QAM modulation this translates into a theoretical bound of 600 Mbps when using two MIMO layers, which if split into 400/200 Mbps (average speed given time multiplexing) equals the capacity need of about 135 passengers. Applying a 50% margin due to coding rates and radio imperfections gives us a 200/100 Mbps realistic data rate to serve 70 passengers, again assuming 15/7.5 Mbps per user and 20% activity factor. For a typical mid-size modern passenger aircraft, 70 passengers in a coach configuration would fit in 12 seat rows, each having a 3+3 arrangement. As a result, the service area per access point can be estimated to stay below a 10 m radius. For smaller aircrafts with narrower fuselage larger service areas can of course be envisaged, thus limited by output power levels rather than capacity constraints.

Fig. 3. Multiple access point installation to achieve needed cabin capacity.

A possible layout is depicted in Figure 3 where the access points are placed in the ceiling above the aisle.

Radio Requirements

In an airplane cabin one can expect large variations of pathloss due to non-line-of-sight conditions. At mmWave frequencies the cabin interior fabric and seat materials used will have significant impact on the reflection properties which makes it hard to derive a channel model without measurements. In addition, blockage due to human bodies and other obstacles will add losses, which can easily exceed 15 dB at these frequencies [11]. Thus, the receiver requirements will have to be based on simple estimates and provide enough margins for unexpected variations. As concluded in the previous section the baseline assumption is an access point located 1 to 10 m away from active devices and mounted in a way that minimizes the average pathloss. Assuming line-of-sight (LoS) for simplicity, the free space pathloss at 39 GHz will range between 64 dB (1 m) to 84 dB (10 m) which yields just 20 dB of variation. This is comfortably within the output power control of mobile device user equipment.

However, given the short distance to secure capacity needs, this system is not anticipated to be sensitivity limited. Instead, we must focus on a different scenario the system will face, which is not when the airplane is in the air but when parked at the gate or taxing to and from the runway. In those situations, the cabin 5G network will coexist with terrestrial services and devices onboard might very well be connected to those. As a result, there are two sources of potential high-level interference presence, i.e., a nearby base station (BS) performing downlink transmission in the direction of the airplane (e.g., serving devices onboard the aircraft), and a device close to the in-cabin access point transmitting to an outside network, as depicted in Figure 4. To understand the dimensioning factors, we consider a worst-case scenario where the wanted device is connected to the in-cabin access point 10 m away. At the same time, another interfering device is located only 1 m away from the access point and connected to a BS (using the same frequency band) in proximity (e.g., 50 m) of the airplane. As aforementioned, the scenario becomes very different during flights as the unsynchronized interferers are removed. In the following sections we will investigate specific requirements for the access point radio to handle the on-the-ground situation. The intention is to stay aligned with the 3GPP NR standard for local access points and user equipment [12, 13], but also to derive target numbers for specific requirements that push performance. It is important to notice that the work is not a product

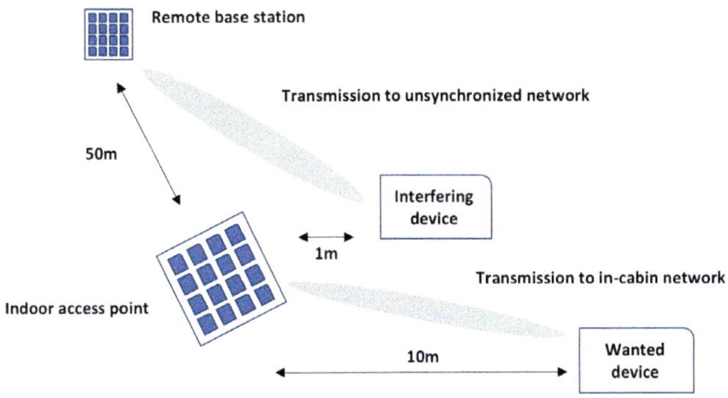

Fig. 4. Radio interference scenario.

specification but serves as a guide for the implementation work in BEYOND5.

Access Point Receiver Requirements

The access point is envisaged to be equipped with 16 cross-polarized antenna patches (i.e., 32 antenna ports) in a 32 TX/32 RX MIMO configuration [14]. This will enable a MIMO gain of 15 dB, plus an additional patch gain of 5 dBi which in total sums to 20 dB. As a starting point, we conclude the fundamental receiver requirements based on the 5G NR standard FR2 n260 class-2 equipment [12], where a device is expected to have a maximum total radiated output power (TRP) equal to 23 dBm. Assuming isotropic radiation and 1 m separation between an interfering device and the access point, the maximum interferer will reach about –36 dBm at each antenna port, as illustrated in Figure 5. The other source of interference comes from the remote BS which is assumed to be a large array system. A 128-antenna-array-system (AAS) producing 10 dBm per antenna port would deliver some 57 dBm EIRP interference in 100 MHz (smallest assumed bandwidth for a wide area deployment). Adding the pathloss (50 m), 5 dBi patch gain, and adjusting to 50 MHz bandwidth, the received interference becomes –36 dBm at each antenna port. Taking penetration losses of the fuselage into account, we conclude that this interference is substantially lower than the close-by interfering device and will not be considered

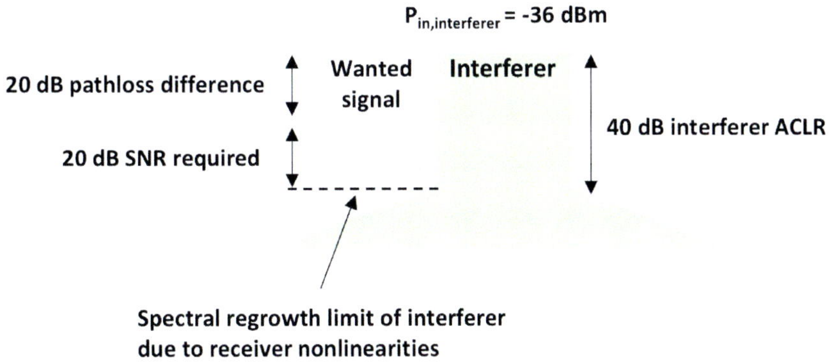

Fig. 5. Signal levels present at the access point antenna ports.

further. As a result, the strongest assumed co-channel interference at each antenna port is –36 dBm. We also conclude that the weakest wanted signal ends up at –56 dBm (10 m distance causing 20 dB additional pathloss) at each antenna port. In a worst-case scenario one would not consider a MIMO gain difference between an interferer and the wanted signal, i.e., assuming the interferer to have the same channel profile as the wanted signal. However, in the targeted airplane cabin deployment this is very unlikely to happen given the anticipated favorable MIMO environment (multipath transmission) and will not be considered. For a 64 QAM modulation with 10^{-5} bit error rate, a detector SNR of 25 dB is required [15] and applying a healthy 5 dB margin for receiver imperfections results in 30 dB. Considering 15dB MIMO gain, this means that each antenna port ideally can suffice with an SNR equal to 15 dB if the noise is fully uncorrelated. However, to compensate for correlated noise and distortion the minimum antenna port SNR is set to 20 dB when exposed to interference.

In the problematic scenario in Figure 4 two mechanisms will contribute to a reduced SNR – the ACLR performance of the transmitting devices and the linearity of the access point receiver. As seen in Figure 5, the third order intermodulation (IM3) distortion creates spectral regrowth of the modulated carrier which falls onto the adjacent wanted signal and raises the noise floor. In addition, the IM3 distortion also limits the signal SNR. Assuming the interferer to have similar performance as the wanted signal (30 dB in-channel SNR), the ACLR is close to 40 dB [16], thus resulting in a co-channel interferer at –76 dBm. This is 20 dB below the wanted signal and in line with the targeted receiver antenna port SNR. The second mechanism has to do with the receiver IM3 distortion and its own spectral expansion of the modulated interferer. We want to dimension the receiver not to hamper the needed SNR in the presence of the adjacent interferer. As the interferer has a different channel profile than the wanted signal (no MIMO gain), maintaining 20 dB SNR is enough and requires a co-channel interferer below –76 dBm (equal to the ACLR contribution). This corresponds to an IIP3 larger than –17 dBm and a 1 dB compression point (CP1) above –27 dBm, which already has significant headroom to the strongest anticipated interferer level (–36 dBm).

If we compare the two sources of co-channel interference it seems reasonable that the ACLR of the interfering device and realistic receiver nonlinearities, causing co-channel noise, could be at the same level. Adding their contributions will then degrade the needed antenna port SNR by 3dB. Consequently, the data rate might be affected as a lower order modulation and coding scheme (MCS) must be employed in combination with a higher bit error rate. As the interfering device is dislocated from the access point, its impact on the system uplink will gradually diminish and at a 2 m distance the co-channel noise contribution is reduced by 6 dB, thus fulfilling the SNR requirement.

To estimate the needed dynamic range in the analog-to-digital converter (ADC) we start with the minimum 20 dB receiver branch SNR. To this we add 20 dB fading margin, 7 dB peak-to-average-power ratio (PAPR) due to OFDM signaling, and 4 dB gain/offset error margin which combined sums to 51 dB signal-to-noise and distortion (SNDR) requirement. This equates to 8.2 effective number of bits (ENOB). To make the quantization noise contribution insignificant, 10 bits resolution is expected to be required.

Access Point Transmitter Requirements

The maximum EIRP cannot exceed 24 dBm for a local area BS [13], meaning

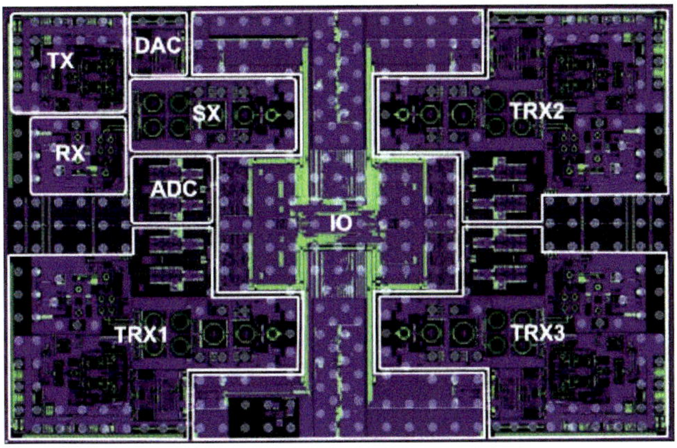

Fig. 6. RFIC chip layout showing the four transceivers and common IO.

that the TRP must take the antenna directivity into account. Given the 32 antenna ports of the access point this results in 4 dBm TRP and –11 dBm modulated TX output power per transmitter branch. Adding anticipated losses in the antenna switch and connectors (4 dB) and the assumed 7 dB PAPR the saturated power of each transmitter power amplifier (PA) becomes 0 dBm. This is about 10 dB lower than the reported CMOS PA power saturation for 39 GHz [17].

To produce a similar momentaneous data rate performance in downlink as for uplink the detector SNR should be the same, i.e., 25 dB as concluded for the access point receiver. Assuming similar margins (30 dB SNR at the receiver antenna port) and limited MIMO gain due to correlation (5 dB), each antenna port on the access point array should have approximately 25dB SNR. This is equivalent to about 6% error vector magnitude (EVM) when summing all transmitter imperfections, as illustrated in Table 1. The different EVM contributions are listed based on realistic performance for the implementation. The PA is expected to dominate when operating close to compression and associated nonlinearities to be strongly correlated in line-of-sight conditions. The image rejection and LO leakage targets assume post-calibration performance, and the antenna mutual coupling is based on single input/output digital pre-distortion

TABLE 1
EVM DISTRIBUTION IN THE ACCESS POINT TRANSMITTER

Imperfection	Value	EVM contribution
PA nonlinearity	Co-channel power ratio = 28 dB	4%
Image (I/Q imbalance)	–35 dBc	1.8%
LO leakage	–40 dBc	1%
Integrated phase noise	1 deg	1.7%
DAC nonlinearity	–30 dB	3%
Antenna mutual coupling	–25 dB	0.3%
Total EVM		$\sqrt{\Sigma EVM^2} = 5.7\%$

(DPD) and can be derived from the caused adjacent channel power leakage [18].

The transmitter is expected to include a third order DPD functionality which will expand the required bandwidth of the transmitter three times, resulting in a baseband bandwidth equal to $3 \times 25 = 75$ MHz. This results in a sample rate equal to 600 MSps when running 4× oversampling. The SNDR is based on the budgeted DAC EVM contribution (3%), the PAPR (7 dB), and a gain control margin (4 dB) which sum to 41 dB (6.6 ENOB). Given implementation margins it is envisaged that each DAC will contain 8 physical bits which results in 9.6 Gbps for each IQ transmitter.

For a local access point, the ACLR minimum requirement is specified to be –20 dBm/MHz [13] which equals –3 dBm for 50 MHz bandwidth. Relative to the maximum TRP it becomes 27 dB. However, as the limiting factor relates to the co-channel power ratio caused by IM3 (similar to the device requirement) the ACLR needs to be approximately 40 dB.

III. DEMONSTRATOR

The scope of the activity has been from the very start to showcase a physical implementation in a real environment. From the use case descriptions and the system requirements, realization requirements have been extrapolated and added in parallel. This has been related to both the transceiver subsystem as described in Section 2 and the digital baseband (DBB), hosting the digital front-end functionality. After thorough analysis, the project decided to use an FPGA development platform for the DBB implementation. The reason was two-fold: focus effort on the actual implementation by reducing the overhead of taping out a digital ASIC and maintaining flexibility throughout the duration of the project rather than having to do a feature freeze a long time ahead of the hardware bring-up. In a similar manner it was concluded that the 22 nm FDSOI process would be the best option for the RFIC as it includes high-power analog, data conversion and a high-speed IO on the same die.

The local oscillator (LO) generation is a frequency tripler which means that the system will have a 13 GHz signal generator output as the input to the RFIC SX block (frequency generator; synthesizer). Furthermore, there is no reference clock generation implemented which means that the system also needs a differential input symbol reference clock driving both the digital baseband clocking and the RFIC IO and the data converter blocks.

In order to perform validation and demonstrations, there are also mechanical requirements which state that the system, apart from off-the-shelf equipment, has to be a single self-contained unit. This is necessary to enable re-positioning of the equipment for OTA measurements in the lab and during in-cabin demonstrations.

System Realization

The activity has multiple partners, each contributing with a functional block design according to their commitments in the project application. Starting at the antenna, two partners are independently developing a dual polarized 4 × 4 antenna array i.e., 32 ports. The footprint for the two different antenna components is similar making it possible to mount either design, or both to cater for efficient evaluation. The antennas modules are 19 mm × 19 mm. Test structures of the antenna have been designed and evaluated by the respective partners along with yet another project partner contributing to the over-the-air testing in the project.

The RFIC is a cooperation between six partners, each responsible for a functional block – SX, RX, TX, ADC, DAC, and a high-speed interface. Each RFIC contains four transceivers and two high-speed interfaces. In addition, there is supporting functionality such as an SPI interface for configuration. The silicon measures 5.9 mm × 3.7 mm. The digital baseband implementation is provided by two of the project partners and makes use of high-performance FPGA development boards.

Building Practice

To be able to measure RF performance, an interposer is being designed to serve as packaging of the RFIC. Two of the transceiver RF ports will be connected to probe pads, and the other two RF ports are connected to BGA balls on the interposer. The interposer also contains de-embedding structures intended for probe measurements. Consequently, the RF evaluation PCB is also being designed to support this. The PCB contains power connectors and high-speed interfaces necessary to perform the RF evaluations. It also complements the interposer probe pads with coaxial connectors for the remaining two RF ports. There are also decoupling capacitors included on the interposer.

The full demonstrator interposer will contain the 32-port antenna module, the eight RFICs to drive the 32 antenna ports, and decoupling capacitors. Unlike the RF evaluation interposer, this interposer will have all RF ports connected to the antenna module. There are no probing or coaxial connectors available so only over-the-air measurements (OTA) will be possible.

The demonstrator system PCB will at its core contain the demonstrator interposer module and the FMC connectors to interface with the digital baseband FPGA platform. The PCB will also implement a 32-way splitter functionality to split the 13 GHz LO input signal to all eight RFICs. This significantly reduces the need for external components in the setup and avoids the problems inherent to such a plethora of cable connections.

The FPGA development board contains, among other features, two FMC connectors, each with access to eight GTH gigabit transceivers [19] and general-purpose I/O signals. Each GTH transceiver will service two transceiver chains, i.e., each RFIC requires two GTH transceivers. Each transceiver will run at 8 Gbps which is approximately half of the theoretical limit of 16.3 Gbps. This design decisions were made to reduce design complexity and reduce the risk.

For lab validation and demonstration purposes, the system board and the DBB FPGA development board will be mounted in an enclosure together with other hardware (e.g., clock splitters, supply cables). This will ensure proper cable management and all-together reduce the mechanical stress on the PCBs. The case will have mounting options for, e.g., a camera tripod allowing for easy re-positioning during lab OTA validation and demonstration campaigns. The overall system is depicted in Figure 7.

Planned Validation

The first part of the system validation is the lab bring-up of the RF evaluation PCB. The RX, TX, and IO partners will have access to the RF-evaluation PCB assembly containing the single-RFIC interposer with probing and coaxial connectivity access. The purpose is to bring up the high-speed interfaces and to test functionality and performance of the RF parts. Following the bring-up is the performance testing on the RF-evaluation board assemblies. All this validation will be performed at each partner site.

Once the full demonstrator system board is assembled and available (containing the module with eight RFICs and an antenna), the bring-up and basic

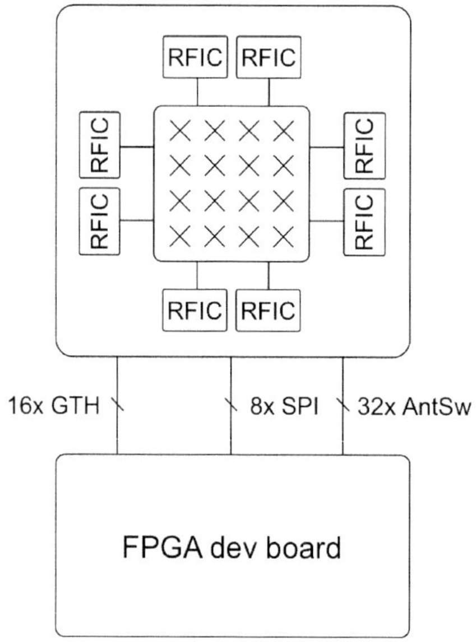

Fig. 7. Demonstrator architecture.

functional tests will precede the first OTA measurements. A remote lab will be set up allowing partners to be connected and contribute to the bring-up, evaluation, demonstrator preparations and support.

The final step in the BEYOND5 project includes an in-cabin demonstration at the Airbus Ottobrun facility where the system will be brought up in an authentic cabin mockup. Demonstrations will be performed that are more relevant to a real deployment, e.g., investigating channel conditions and performance of the system by testing different positions for both base stations and user equipment. Other demonstrations might be added but are currently not being prepared for as part of BEYOND5.

IV. Conclusion

In this paper we have described the basic conditions for mmWave deployments and the market outlook as a background to the demonstrator work performed in the ECSEL-2019 project BEYOND5. Initial thinking and derivation of selected radio requirements were presented given the targeted use case. We also covered the demonstrator planning and the associated engineering work needed for the system bring-up and final OTA measurements in a real environment. The activity is planned to continue until mid-2024 and follow the regular dissemination plan of the BEYOND5 project.

References

[1] GSA. [Online], 2022, November. Available: http://gsacom.com/paper/nts-statistics-november-2022/

[2] Ericsson, *Ericsson Mobility Report* [Online], 2022, November. Available: http://www.ericsson.com/mobility-report/

[3] B. Sadhu et al., "A 28-GHz 32-element TRX phased-array IC with concurrent dual-polarized operation and orthogonal phase and gain control for 5G communications", *IEEE Journal of Solid-State Circuits*, vol. 52., no. 12, pp. 3373-3391, December 2017.

[4] M. Shafi et al., "5G: A tutorial overview of standards, trials, challenges, deployment, and practice," *IEEE J. Sel. Areas Communication*, vol. 35, pp. 1201–1221, June 2017.

[5] 3GPP, *Ultra-Reliable and Low Latency Communication* [Online], 2022, January. Available: https://www.3gpp.org/technologies/urlcc-2022/

[6] Ericsson, *6G Spectrum - Enabling the Future Mobile Life Beyond 2030* [Online], 2023, February. Available: https://www.ericsson.com/en/reports-and-papers/white-papers/

[7] BEYOND5 H2020-ECSEL-2019-1-IA. [Online]. Available: https://www.kdt-ju.europa.eu/projects/beyond5

[8] G. Mangraviti et al., "A 39-GHz 18.5-mW LNA with T/R switch, 15.4-dB gain, –2.2dBm IIP3, 5.6-dB NF, for a 5G in-cabin base station in 22-nm FD-SOI", *International Microwave and Radar Conference, Gdansk, Poland, September 2022.*

[9] K. Trzebiatowski et al., "Multibeam antenna for Ka-band CubeSat connec-

tivity using 3-D printed lens and antenna array," *IEEE Antennas and Wireless Propagation Letters*, vol. 21, no. 11, pp. 2244-2248, November 2022.

[10] NGMN Alliance, *5G White Paper* [Online]. 2015, February. Available: https://www.ngmn.org/work-programme/5g-white-paper.html

[11] Q. Wang et al., "Attenuation by a human body and trees as well as material penetration loss in 26 and 39 GHz millimeter wave bands", *Hindawi International Journal of Antennas and Propagation*, vol. 2017, pp. 1-8, 2017.

[12] 3GPP (December 2019), *TS 38.101-2 v15 8.0*. [Online]. Available: https://www.3gpp.org/specifications-technologies/releases/release-16

[13] 3GPP (December 2019), *TS 38.104 v16 2.0* [Online]. Available: https://www.3gpp.org/specifications-technologies/releases/release-16

[14] K. Trzebiatowski et al., "A dual-polarized 39 GHz 4x4 microstrip antenna array for 5G MU-MIMO airflight cabin connectivity", *24th International Microwave and Radar Conference (MIKON), Gdansk, Poland, 2022*

[15] D. Coudert et al., "Wireless backhaul networks: minimizing energy consumption by power-efficient radio links configuration", January 2008 [Online]. Available: https://www.researchgate.net/

[16] N. Borges de Carvalho et al., "Compact formulas to relate ACPR and NPR to two-tone IMR and IP3," *Microwave Journal*, December 1999.

[17] J. Park et al., "A 26-to-39GHz broadband ultra-compact high-linearity switchless hybrid N/PMOS bi-directional PA/LNA front-end for multi-band 5G large-scaled MIMO system," *IEEE International Solid-State Circuits Conference (ISSCC), San Francisco, USA, February 2022.*

[18] X. Chen et al., "A review of mutual coupling in MIMO systems," *IEEE Access*, vol. 6, pp. 24706-24719, 2018.

[19] Xilinx High Speed Serial Technologies, [Online] 2023. Available: https://www.xilinx.com/products/technology/high-speed-serial.html

Fredrik Tillman received his M.Sc. and Ph.D. degrees in Electrical Engineering from Lund University in 2000 and 2005 respectively. After graduation he joined Ericsson Mobile Platforms and participated in the first cellular modem CMOS radio development in Sweden and US before moving on to the research branch of the company. Today he is heading a department at Ericsson Research with focus on integrated radio circuit design for both cellular infrastructure and device connectivity. Besides being responsible for internal R&D activities, Fredrik is active in the European research community and has been the Ericsson driver for multiple collaboration projects within the Horizon 2020 framework. He is currently the Technical Manager for the COREnext project. Since 2022, Fredrik has served as a board member of the Swedish strategic innovation program Smarter Electronic Systems to increase competitiveness and growth in Swedish industry.

Pallavi Paliwal received her B.E. degree in electronics and communication engineering from Gujarat University, India, in 2006, and her M.Tech. and Ph.D. degrees from IIT Bombay, India, in 2018. From 2006 to 2008, she worked with eInfochips, India, as an ASIC design engineer. She is presently a senior researcher with Ericsson, Lund, Sweden. Her current work experience includes frequency synthesizers and transceiver systems.

Daniel Eckerbert received his M.Sc. from Linköping University in 1998 and his Ph.D. from Chalmers University of Technology in 2003 respectively. During his Ph.D. studies, he spent half a year internship at Intel's Microprocessor Research Lab in Hillsboro, Oregon, working on multi-phase clock generation. After graduation he joined Imego, working on system design and project management for MEMS inertial measurement unit systems. In 2007 he joined Ericsson Research in Lund and is currently a Senior Specialist in radio architecture and integration. Daniel's current research focus is on integration topics as they relate to integrated radio system design.

Digital Beamforming Transceiver Design in 22 nm FD-SOI Technology for 39 GHz 5G Access

J. Prouvée, G. Mangraviti, B. Debaillie, P. Wambacq,
D. Borggreve, R. Ciocoveanu,
H. Fredriksson, P. Paliwal, F. Tillman, H. Terlemez, B. Dündar,
V. Pinon, F. Hasbani, A. Ferret, C. Dehos,
A. Hamani, B. Martineau, and D. Morche

Abstract—**Pure digital beamforming is the best strategy to address dense end-user area. However, this architecture comes with its own challenges (power consumption, extra area, extremely high I/O data rates) which often makes it considered to be a non-practical solution for antenna array integration. In this paper, we will discuss how these inherent challenges have been addressed in the development of fully integrated transceiver solution operating at 39 GHz. Low power RF front-end with Rx/Tx switching solution will be presented. Local oscillator multiplication from 13 to 39 GHz will be detailed. Efficient and high resolution data conversion will be justified and presented. Lastly, a new architectural solution addressing the issue of high-data rate transfer will be explained.**

Index Terms—**Bandpass filter (BPF), base station (BS), digital beamforming (DBF), digital pre-distortion (DPD), low noise amplifier (LNA), low pass filter (LPF), power amplifiers (PA)radio-frequency integrated circuit (RFIC), serial peripheral interface (SPI), successive approximation analog-to-digital converter (SAR ADC).**

The research leading to these results has received funding from the European Union's ECSEL Joint Undertaking with finanical support form France, Germany, Sweden, Belgium, Poland, Netherlands, Romania, Turkey, Israel, and Switzerland, under grant agreement n° 876124 – project BEYOND5. The authors thank Globalfoundries Dresden for the support provided for the MPW manufacturing, and Ericsson for its helpful guidance in the system specification and work package animation.

J. Prouvée, C. Dehos, B. Martineau, and A. Hamani, D. Morche are with CEA-Leti France. G. Mangraviti, B. Debaillie, and P. Wambacq are with imec, Belgium. D. Borggreve is with Fraunhofer EMFT, Munich, Germany. R. Ciocoveanu, was with Fraunhofer EMFT, Munich, Germany. He is now with NXP Semiconductors. H. Fredriksson, P. Paliwal, and F. Tillman are with Ericsson, Sweden. H. Terlemez, and B. Dündar are with MKR Turkei. V. Pinon, F. Hasbani, and A. Ferret are with ASYGN France.

I. INTRODUCTION

THE spectral crowding of the FR1 frequency range (410 MHz to 7.125 GHz), combined with the need for increased low-latency and high data rate links, has made the FR2 frequency range (24.250–71 GHz) an obvious candidate for 5G and beyond communications. The inherent challenges to work in the millimeter wave (mmW) domain itself are major (high free-space loss, rain impact, foliage loss, high penetration loss of new construction materials, human shadowing), even if benefits are rewarding [1][2].

Beamforming using massive MIMO (multiple input multiple output) is an interesting strategy to counterweight some of those mmW limitations. The directivity brought by the *n* antenna array improves the budget link by a factor 10 × log(n) on each side of the transceiver. Phase shifting those signals allows beam steering that permits spatial selectivity, game changing the networks regarding interferences and shadowing [3]. From there, two transceiver architectures can be used: digital or analog (plus hybrid) beamforming, respectively DBF or ABF. Figure 1 illustrates both approaches at the transmitter level. Analog beamforming consist of phase shifting the same signal on each antenna (thus steering a single beam). With digital beamforming, phase shifting is performed digitally with a higher degree of freedom, generating single or multiple beams, spatial filtering, etc.; hybrid beamforming combines both techniques [2][4].

Many works have demonstrated notable performance on both analog and digital architectures, summarized in Table 1 (note that the power consumption of this work is detailed in the conclusion).

Analog beamforming systems generally achieve performances with lower power consumption, considering RF front-end and digital base-band (DBB). Its number of paths is usually limited due to the difficulty of matching and synchronizing a large number of RF components. On the other hand, DBF (digital beamforming) opens several opportunities regarding RFIC design:

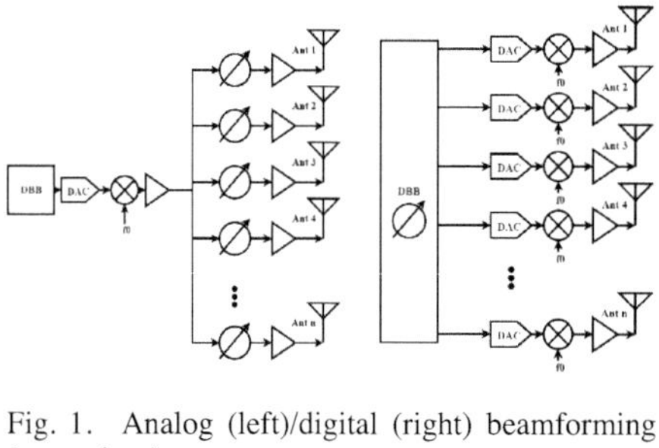

Fig. 1. Analog (left)/digital (right) beamforming (transmitter).

- DBF allows transmitting or receiving multiple beams simultaneously.
- This architecture is widely scalable (number of antennas can easily vary without need for redesign), making it an ideal candidate for massive MIMO.
- Parallelizing a large number of low power PA (power amplifiers) and making over-the-air recombination is more efficient than on-chip power recombination in high power PA.
- Amplitude weighting of each path allows minimizing the side lobes of the antenna beams. DPD (digital predistortion) can be differentiated on each path, leading to better linearity and smaller power back-off.

That being said, RFIC design for DBF still has major challenges to face:

- Being able to increase drastically the number of antennas thanks to scalability means that RF power consumption can rapidly take considerable proportions, even if the implementation losses are less predominant compare to analog beamforming (power combining and phase shifting).
- The dynamic range of the RF front ends has to be very large in order to take advantage of the flexibility/scalable properties of DBF.
- DBF requires high frequency LO distribution with a perfect synchronization, which is definitely more difficult to manage at mmWave frequencies.
- The volume of data on I/Os can take huge proportions. Consider our work as an example that leads at the transmitter side of each of the 8 RFIC to an I/O data rate of 47.2 Gbps: 8-bit DAC @737.28 MSps × 2 (I&Q path) × 4 (Nbr of channels per RFIC).

TABLE 1
RECENT MMWAVE BEAMFORMING RFICS

Ref	Beam-forming	Frequency (GHz)	Area (mm^2)	Architecture	Channel per IC	Technology	RX Pdc (mW)	TX Pdc (mW)
[5]	Analog	24.2–30.5	17	TRX	4	130 nm SiGe BiC-MOS	1600	1800
[6]	Analog	37–40	17.2	TRX	8	28 nm RF-CMOS	78.5/ch	339/ch
[7]	Digital	52.5	3.3	RX+AD-C+Dig Beamformer	4	28 nm RF-CMOS	96/ch + 372	/
[8]	Hybrid	25–30	3.86	RX	8	65 nm CMOS	340	/
[9]	Analog	26.5–29.5	30.08	TX	16	28 nm CMOS	/	1630
[10]	Digital	28	5.76	RX+Mux	4	65 nm CMOS	60/ch	/
[11]	Analog	62	441	TRX+ADC/DAC	64	180 nm SiGe BiC-MOS	4500	5125
[12]	Analog	17.1–52.4	12.5	TX	4	180 nm SiGe BiC-MOS	/	240/ch
[13]	Digital	28	7.73	RX+IO	16	40 nm CMOS	2800	/
[14]	Analog	39	30 + 33.4	TRX+IF transc.	16	28 nm + 65 nm CMOS	624	1680/1840
This work	Digital	39	22.42	TRX+ADC/DAC+IO	4	22nm FD-SOI CMOS	461	843

The following sections will explain a practical approach and a specific RFIC case of how to address these challenges. The use case is an in-cabin 5G base station (BS) [15]. The distance between the UE and BS is between 1 and 10 m. The data rate has been determined using the Next Generation Mobile Network (NGMN) Alliance proposal [16]. Since expected users' experienced data rate is meant to reach 15/(7.5) Mbps [downlink/(uplink)], a 1.2 Gbps/(600 Mbps) downlink/(uplink) speed is required for the whole cabin system (assuming a 20% activity factor with 400 users in the plane).

Some of the following secions will present measured data, some will only focus on simulation results. This is due to the fact that at first silicon was manufactured and allowed preliminary measurements on building blocks. The full RFIC (Figure 2) has been manufactured but has not yet been measured.

II. ARCHITECTURE

A four-path transceiver architecture is proposed as a trade-off between complexity, space occupancy and integration, and I/O data rate. Each chip is

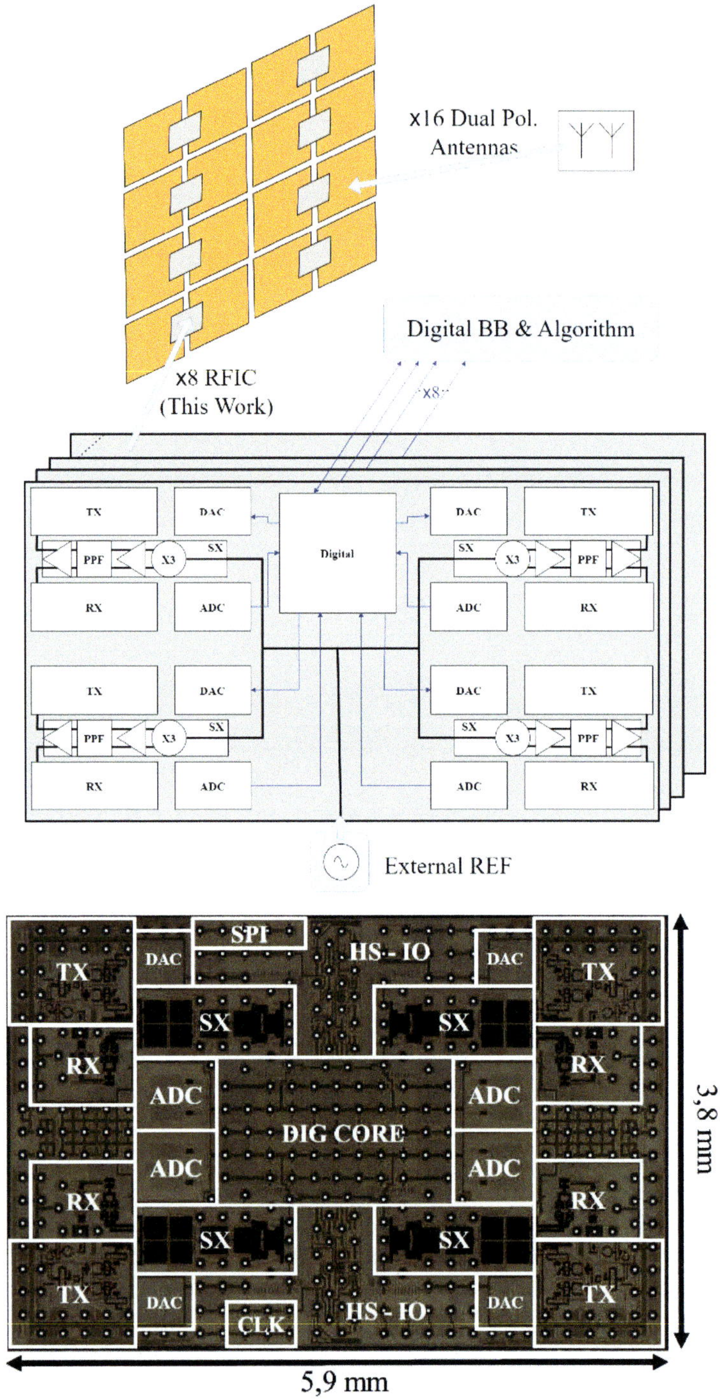

Fig. 2. DBF system and RF chip architecture (top), chip photograph (bottom).

intended to feed a two-port dual polar antenna (see Figure 2). At 39 GHz, the $\lambda/2$ antenna pitch is 3.8 mm. In our use case, eight ICs of four transceivers are expected to match the requirements of a small Class 2-0 base station (BS) radio transmission and reception operating in FR2 n260 channel (37–40 GHz), reaching a maximum of 24 dBm EIRP at the transmitter side. The link budget gives a net data rate of 150 Mbps per 50 MHz channel at 10 m range in downlink, between BS and UE. The transceiver can achieve up to 225 Mbps at closer distance, using OFDM-64QAM modulation. Longer distance and higher data rates can be obtained scaling up the array size and number of chips, increasing EIRP.

The PA output power is willingly limited (8 dBm P1 dB) to achieve the better power added efficiency (PAE) at power back-off, with DPD, meeting the standard linearity requirements.

A common reference clock feeds the different ICs. Its frequency has been chosen to be one third of the final LO to reduce the power consumption. Thus, internal frequency multipliers generate the required LOs coherently. The channels (whose bandwidth can scale from 50 to 400 MHz) are up-converted from 37 to 40 GHz by sliding the reference clock between 12.33 and 13.33 GHz. Assuming digital common phase error correction and pilot phase tracking in DBB OFDM signal processing, the challenge is to limit the phase noise floor at frequencies higher than 1 MHz from the carrier in frequency multipliers.

The receiver manages high dynamic range, from –86 dBm at sensitivity level in QPSK modulation, up to –28 dBm at the shortest distance. This is carried out by dynamic gain reconfiguration in low noise and intermediate amplifiers as well as automatic gain control in the variable gain amplifier.

III. RECEIVER

The digital-beamforming base station envisioned in this work challenges the receiver (Rx) chain mostly in terms of linearity, low power consumption and sharing of one antenna port with the transmitter. The linearity and noise figure (NF) are derived based on the user equipment (UE) Tx specifications defined in [17] and a link budget. Considering mainly two scenarios where the distance between the near-end and far-end UE can vary between 1 m and 10 m, the Rx NF and the input-referred third-order intercept (IIP3) result in 7 dB and –10 dBm respectively. Moreover, an even larger linearity (IIP3 well above 0 dBm) is needed for the transmitter digital-predistortion (Tx-DPD), where the Rx chain is used as a feedback loop. The requirement of low power consumption is directly translated into limiting the Rx supply voltage up to the core voltage, i.e. 0.8 V. This challenges the tradeoff between large IIP3 and low power consumption. The requirement on sharing the same antenna port with the transmitter means that a transmit/receive (T/R) switch is needed at the Rx input, to protect the Rx when Tx is on and, at the same time, to still be compatible with ESD reliability.

The Rx chain is drawn in Figure 3 from the antenna T/R interface to the baseband (BB) to a first stage of variable-gain amplifiers (VGAs). It consists of:

- An input matching network interfacing the antenna, the Tx (meaning PA output) and the Rx (meaning LNA input); this includes a switch to se-

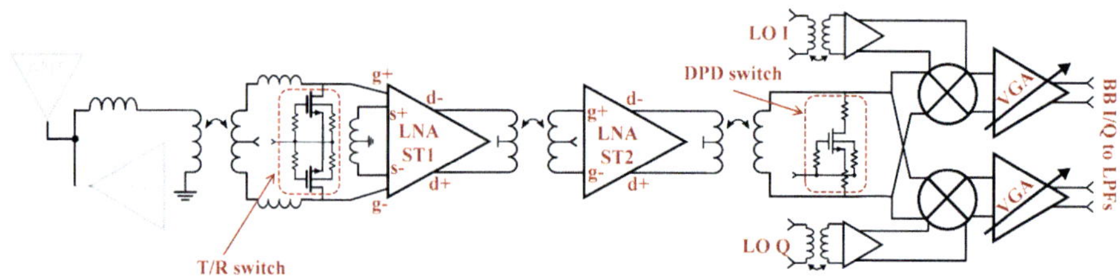

Fig. 3. Receiver architecture from the antenna T/R interface to the BB VGAs.

lect between Tx and Rx mode of operation.

- A two-stage LNA.
- Two differential down-conversion mixers (for I/Q operation).
- Two differential LO amplifiers buffering the I/Q LO signal provided by the on-chip LO multiplier.
- Two differential BB VGAs.

The two-stage LNA, including the input T/R matching network, consists of a pseudodifferential topology (depicted in Figure 3), operating under the core supply voltage. Its design and experimental validation are already reported in [18]. The first stage is a pseudo-differential common source with source degeneration and cross-coupled capacitive neutralization (to boost linearity and simultaneous matching for power and noise). The second stage is a pseudo-differential common source, with cross-coupled capacitive neutralization, designed for large linearity. The differential operation is preferred to single-ended because of better linearity and design robustness against common-mode current-return paths. The matching transformers allow for a compact layout of matching and decouple the DC biases. The transformer in the T/R input matching network protects also against ESD.

This LNA (including input T/R matching) has been already tested standalone [18]. It consumes 18.5 mW under 0.8 V within a core area of 0.1 mm^2. It outperforms the linearity specifications with an IIP3 of –2.2 dBm in Rx mode and +14.9 dBm in Tx-DPD mode. In Rx mode it provides a gain of 15.4 dB at 39 GHz, noise figure below 6 dB, while gain and consumption are reduced in the Tx-DPD mode to –8.6 dB and 8.8 mW.

The two differential down-conversion mixers are passive, double balanced and operate in voltage mode. This circuit topology allows for large linearity trading off gain. These mixers are preceded by a switched differential resistor and a differential splitter. The switched resistor is enabled in Tx-DPD mode, to reduce Rx gain and boost linearity. The FET mixer switches are driven by an I/Q LO signal buffered by pseudo-differential common-source amplifiers with cross-coupled capacitive neutralization and transformer matching. From simulations, each of these LO amplifiers consumes around 3 mW and can work with a differential input power of –8 dBm.

The two differential BB VGAs buffer the down-converted signal from the mixers to the BB low-pass filters (LPFs). As in Figure 4, they consist of pseudo-differential common-source (NMOS) gm-R amplifiers, with single-ended FET (PMOS) load and differential variable resistive load. Here linearity is traded off with gain. A common-mode feedback loop, through the FET load, fixes the output DC operating

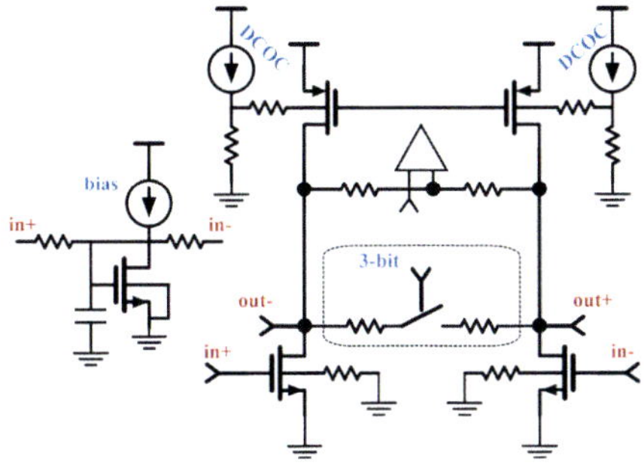

Fig. 4. BB VGA schematic.

point to 0.4 V (as needed by the BB LPFs). The backgates of the FET loads are separately driven for DC-offset compensation (DCOC), which is critical for direct-conversion receiver architectures. From simulations, each VGA consumes 4 mW and can range the gain from 1.7 to 13 dB.

The whole Rx chain (Figure 3), in the nominal Rx mode, exhibits in simulation the performance reported in Table 2, consuming 36 mW under 0.8 V. In Tx-DPD mode the Rx gain is reduced – and consequently linearity further boosted – setting both switches on (T/R switch and DPD switch) and reducing the current bias of the two LNA stages. When this is simulated, the Rx gain can vary between –16.6 and –5.5 dB, with an IIP3 of around +13 dBm, consuming 22 mW under 0.8 V.

TABLE 2
PERFORMANCE IN RX MODE

VGA gain setting	Min	Nom	Max
Rx gain (dB)	17.6	23.9	28.8
VGA gain (dB)	1.7	8.1	13.0
Rx NF (dB)	7.1	7.1	7.1
IIP3 (dBm)	–9.0	–9.8	–12.1

IV. SAR ADC DESIGN

The design of high-performance analog-to-digital converters (ADCs) is of critical importance in many modern electronic systems, including wireless communication and digital signal processing. In this section, we present a SAR ADC design that addresses the challenges of reducing energy loss from capacitive switching, mitigating signal-dependent charge injection, and increasing the accuracy of reference voltage settling time. Our proposed design employs set-and-down conversion, bootstrapped sampling switches, split-capacitor redundancy, and an asynchronous control circuit to achieve high precision and speed performance. We describe the design principles and operation details of our proposed SAR ADC, highlighting its advantages and potential for practical applications in RF integrated circuits.

Set-and-down conversion is a more efficient analog-to-digital conversion technique that requires fewer capacitors and less capacitive switching, thus reducing energy loss from capacitive switching. The set-and-down sequence starts with the MSB comparison before any capacitive switching. Depending on the MSB comparison $V_{ref}/2$ is either added or subtracted from CDAC voltage. Only $N - 1$ capacitors are required for N bits, reducing both area and switching energy. It is also faster as the MSB conversion can be initiated directly after sampling, providing an important timing advantage as settling of capacitive switching is often the limiting factor for speed.

However, there are two potential disadvantages of set-and-down conversion, namely, signal-dependent charge injection from the sampling switch, which can cause distortion and the varying common mode of the DAC, which can impact the precision and speed performance of the comparator. The design uses bootstrapped sampling switches at the inputs that receive boosted gate voltages to have constant on resistance and to mitigate the signal-dependent charge injection, improving linearity.

In SAR ADCs, the reference voltages are used to generate comparison points for conversion steps. The reference voltage settling time is important because it directly affects the accuracy of the ADC. Our design employs the split-capacitor redundancy technique that splits the original MSB capacitor into several sub-capacitors that are either added to or inserted between the original binary capacitors with each additional split-capacitor increasing the total number of conversion steps. This results in overlapped searching paths and extended search ranges for LSB steps, while the total value of CDAC capacitors remains the same. Our design utilizes three additional cycles per sample to provide sufficient redundancy to cover the CDAC settling error. The split-capacitor redundancy technique uses only adders for its digital error correction logic.

The proposed ADC (Figure 5) employs an asynchronous control circuit to internally generate the necessary clock signals, avoiding the need for a high-frequency clock signal. The operation period of the ADC is divided between sampling and conversion phases, with a ratio of 25–75%. The sample and convert signals are generated from a clock source that is four times faster than the sampling rate. The ADC starts comparisons at the end of sampling. The asynchronous control circuit uses a ready detector at the comparator output to generate a valid signal, eliminating waiting time caused by meta-stability problems. The valid signal is used to generate a trigger signal for SAR logic and a compare/reset signal for the comparator. Then, the valid signal initiates the comparator reset phase, coinciding with CDAC settling time, with its duration controlled by an adjustable delay circuit. The SAR logic generates an end of conversion signal after all conversion steps are completed. The background calibration circuit

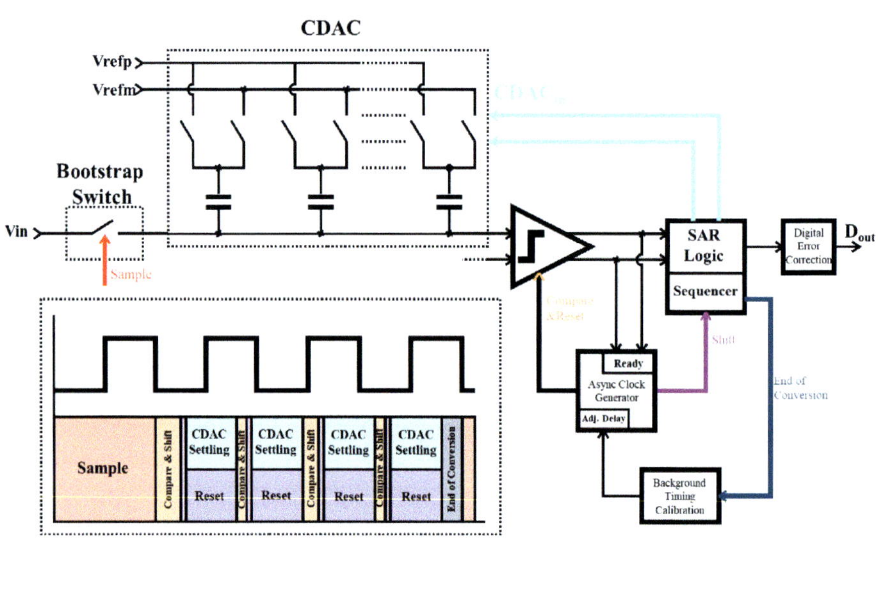

Fig. 5. SAR ADC block diagram.The proposed ADC

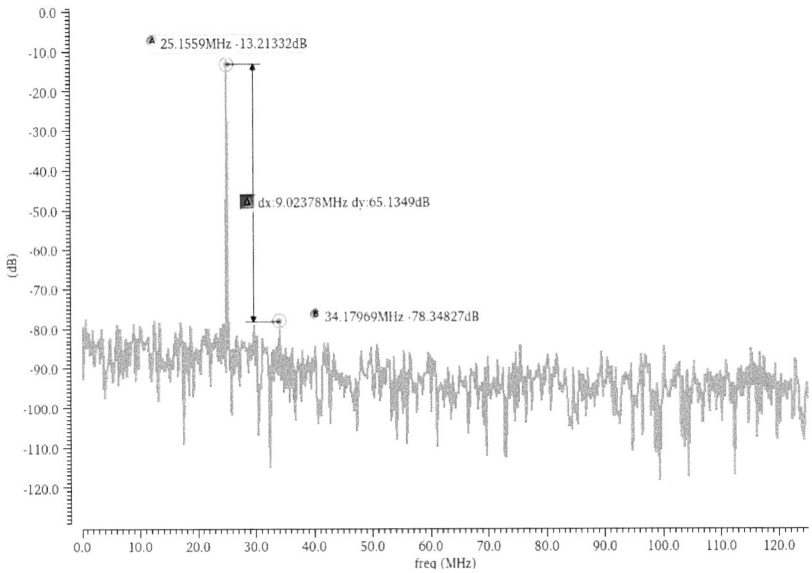

Fig. 6. Output spectrum for 25.1MHz input signal sampled at 250 msps.

adjusts the delay to allocate more time to CDAC settling and minimize the idle time of the ADC.

Figure 6 plots the FFT spectrum for an input frequency of 25.1 MHz, while the ADC operating at a sampling rate of 250 msps. The ADC exhibits 49.6 dB signal to noise and distortion ratio (SNDR) performance, resulting in an effective number of bits (ENOB) value of 8 bits, and 65 dB spurious-free dynamic range (SFDR) performance.

In conclusion, the SAR ADC design presented in this study showcases good performance in terms of efficiency, accuracy, and speed. The implementation demonstrates impressive operational capabilities, achieving a sampling speed of 250 msps while consuming less than 29.6 mW of power.

V. TRANSMITTER

In this section, we present the design of the Tx circuit part shown in Figure 7. The overall assembly of the Tx includes two baseband amplifiers, two mixers with an LO buffer, two variable amplifiers for both I and Q paths before combining, and a power amplifier. The I /Q paths are combined to provide one differential signal to the power amplifier. This architecture allows a dedicated gain control for the I and Q path in order to compensate for gain variations and IQ mismatch between paths. In addition, an original LO leakage control using FDSOI transistor back gate characteristics has been used and can be digitally adjusted. A serial peripheral interface (SPI) allows digital control and full reconfiguration of the Tx settings, to guarantee per-

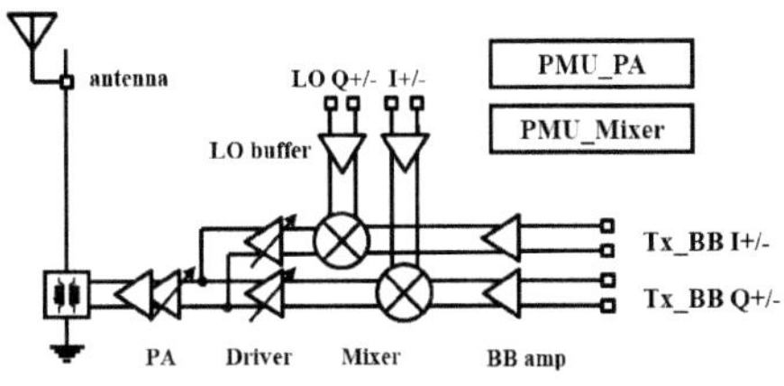

Fig. 7. Tx architecture.

fect frequency synchronization and gain matching of the different paths. It includes also its own power management units for better Tx/Rx isolation and immunity to supply noise, and a dedicated analog bus to sense the voltage/current nodes, output power at mixer and PA with the aim of test and digital calibration of the Tx. Finally, it includes squarer based power sensing after the mixer and the power amplifier.

Power Management Unit (PMU)

The PMU (represented Figure 8) has been design to provide voltage reference, linear voltage regulation, and current reference. It is composed of a temperature and voltage supply independent voltage reference, a low drop out voltage regulator (LDO), a proportional to absolute temperature current reference (IPTAT) and a programmable current DAC. The 6 bits DAC allows controlling each bloc biasing circuitry from programmable registers thanks to the SPI. A low voltage supply is used in accordance with the requirement of 22FDSOI technology (i.e. <1 V), for low power consumption. It is based on the bandgap reference presented in [19]. Instead of a parasitic bipolar transistor at the output, it uses an NMOS transistor in weak inversion. The voltage reference is adjustable from 580 mV to 705 mV thanks to 3 bits to compensate for process variation. The LDO is composed of an op-amp-based voltage regulator and sized to provide up to 100 mA under 0.8 V from a 1.2 V power supply. The output voltage is equal to the reference voltage applied to the non-inverting input of the op amp. This value can be changed from 750 mV to 1 V thanks to a 3 bits programmable resistive divider.

Base-band Amplifier

The base-band circuit consists of an active first-order programmable filter with a power consumption of 0.554 mW (in typical conditions). The input impedance of the baseband amplifier is set to ensure the required voltage gain with a voltage

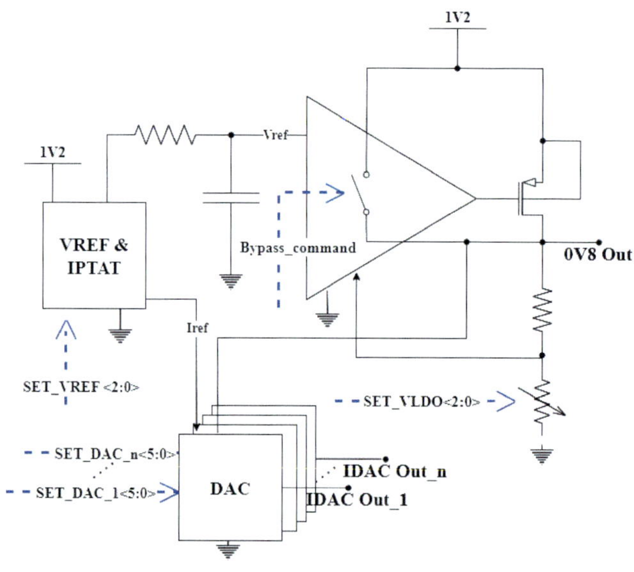

Fig. 8. PMU architecture.

input from the DAC. In addition, a programmable second order filter function is performed. It allows to filter out replica from DAC with a tunable cut-off frequency, that corresponds to three times the half-channel bandwidth (assuming I,Q + DPD).

Up-converter Mixer

The architecture of the designed mixer is based on the Gilbert cell. As shown in Figure 9, bleeding PMOS (P1 and P2) transistors are used to improve the linearity of the up-converter mixer. For this circuit, the length of all transistor is 20 nm, while their widths are set as: N1 and N2 NMOS widths are 24 μm, P1 and P2 PMOS widths are 36 μm, SW1, SW2, SW3 and SW4 NMOS widths are 12 μm. The power consumption of the designed up-converter mixer is 0.54 mW. The back-gate of the transistors are used to control the LO leakage and compensate the IQ mismatch. The LO buffer achieves a single-ended to differential conversion and the interconnections between the LO outputs (from the Sx) and the LO buffer and balanced to reduce the insertion losses and ensure the matching between these blocks.

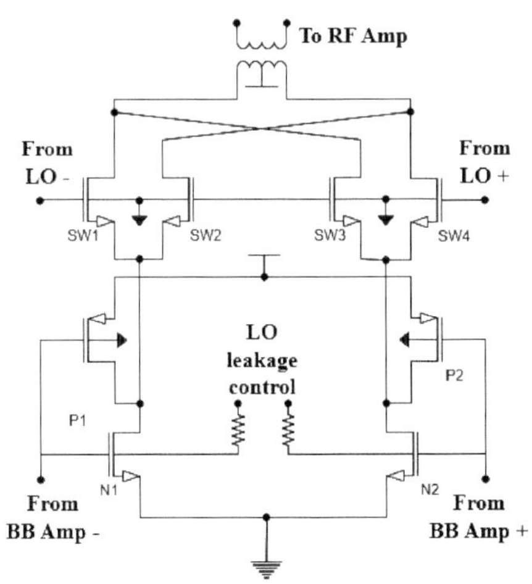

Fig. 9. Up-converter mixer architecture.

Mixer Driver

A variable gain amplifier is designed after each up-converter mixer with six control bits from the SPI. It consists of one stage cascade amplifier with switchable transistors. The variable gain amplifiers allow the control of the gain variation between the I and Q paths and can provide an additional gain up to 10 dB for the overall Tx.

Power Amplifier

The power amplifier includes two stages, a first driver stage with gain control to adjust the Tx gain to correct for any process variations or antenna path mismatch and a second power stage to deliver the required output power. The input and output impedances of the VGA should remain relatively constant, which would otherwise change the impedance seen by the former or latter circuit stages in the channel and introduce additional gain variations. To overcome such an issue, the gain control function uses a current steering technique because of its inherent constant return losses and its digitally assist capabilities [20].

Figure 10 shows the schematic of the digital-controlled VGA with transformer-based baluns at input and output. The VGA consists of differential cascode amplifier and digital-controlled common-gate amplifier array for gain tuning. At maximum gain mode, the total width of the common-gate stage equals to the total width of the common-source stage. The proposed CMOS VGA achieves 4-bit gain control range from 0 to 12 dB at 37 GHz.

Both VGA and power stages use Super Low Vt (SLVT) thin-oxide transistors offering the best fmax performances.

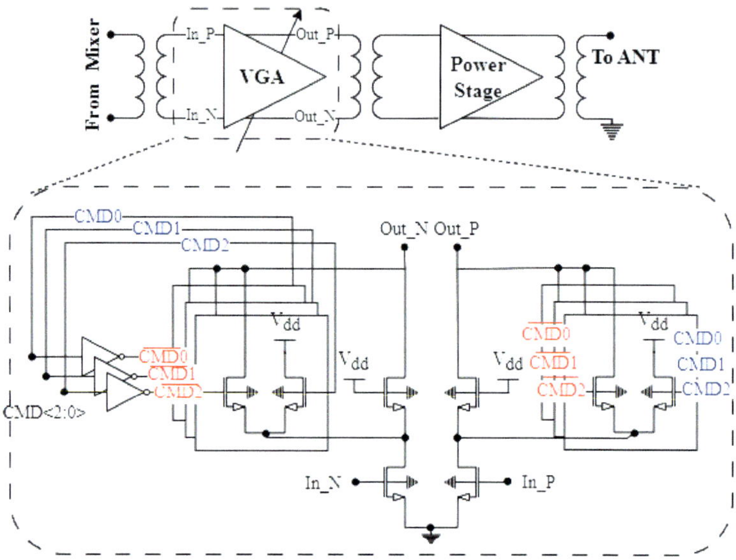

Fig. 10. Power amplifier architecture and VGA first stage.

Thanks to the differential architecture used, the common source transistors can be stabilized thanks to a miller effect neutralization. In this work, PMOS transistors are used for neutralization and results in the cancellation of AM-PM distortion. The PA is centered at 37 GHz with a –3dB bandwidth from 31.8 to 41.8 GHz. The PA's optimum load has been anticipated for an extra output capacitance. Indeed, to account for parasitic capacitances due to the balling and module assembly, the PA has been designed to sustain up to 100 fF extra capacitance with constant performances of 8 dBm output compression point. The power consumption of the PA in nominal conditions is 47 mA per channel and associated PAE is 11%.

VI. TX DAC DESIGN AND VALIDATION

The Tx system linearity requirements imply minimum speed and linearity figures for the modulation DACs. The 50 MHz signal occupies 61 MHz after OFDM CP-extension; a six times higher sampling rate is needed to compensate third order non-linearity, so the required sampling frequency is 368 MHz. To account for new technology unknowns, and optionally improve alias filtering, we pushed the design up to 700 MHz, and could even test it up to 1 GHz. The SFDR and IM3 should be above 50 dBc. It is achieved with 8 bits control and a dynamic element matching (DEM) system.

Figure 11 details the DAC circuit architecture.

The current reference for the converter is generated internally, from a bandgap voltage applied over a poly resistor. It can be scaled over a 36 dB range, acting as a gain control input. Special care is taken for value spread: <2% standard deviation over process voltage and temperature (PVT) + mismatch and power supply rejection ratio (PSRR) (>14 dB from DC to 1 GHz).

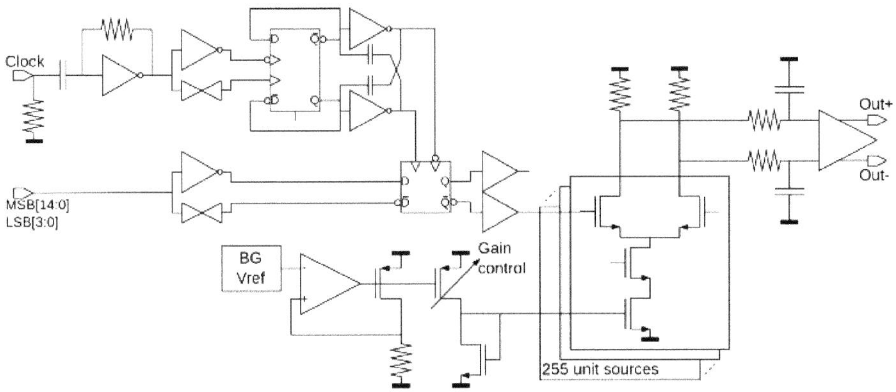

Fig. 11. Tx DAC simplified schematic.

The sampling clock is derived from an AC-coupled double-frequency base-band clock (1.4 GHz typ), and broadcast to the 255 unit switches by differential low-jitter balanced buffers, achieving jitter and imbalance of respectively 0.4 ps and 0.3° typical (0.7 ps and 4.8° max).

The converter's core is based on a differential current steering NMOS structure. Four LSBs are driven by a binary code, while Four MSBs use a thermometric control with DEM scrambling (modulation by a 32 bits Fibonacci LFSR). The unit cell consists of an NMOS current source, with cascode and differential switches, with neutralization devices: always-pass gates controlled by opposite signal to cancel charge injection, with dummy structures balancing all control paths.

The current-steering structure is loaded by a 100 Ω differential load (poly resistor matched to the reference generator), followed by a programmable 95 MHz RC filter. The output is buffered by source followers.

The whole converter is simulated with mismatch in terms of INL and DNL (<0.3 LSB at max gain), SFDR and IMD3 with DEM pattern (62 dB typ, 59 dB worst case, with 10 MHz and 11 MHz tones).

The converter consumes typically 2.6 mA from a 0.8 V supply excluding the output buffer that was sized for stand-alone test purpose.

The layout of the DAC is driven by matching considerations to meet linearity target (INL, DNL, IM3). In advanced nodes and especially in FDSOI, matching characterizations and models indicate that increasing unit device dimensions beyond a few micrometers does not improve performance. For that reason, we chose an NMOS finger pattern of 0.6 μm/2 μm, to interleave devices for Vgs reference, LSB and MSB over a distance of less than 3 μm. Compactness and pattern uniformity also matter much for high-speed operation: we checked that the delays introduced by layout parasitics are balanced, and do not degrade the converted signal.

Finally, the fabricated chip embedded a two-tone direct digital synthesizer. The most relevant measured result for this work is shown on Figure 12, exhibiting the IM3 performance enabling or not the DEM system. The 56 dBc linearity is comparable to the 57 dBc given by transient simulations with mismatch and DEM, confirming the chosen interleaved extended structure took advantage of FD-SOI matching specificities.

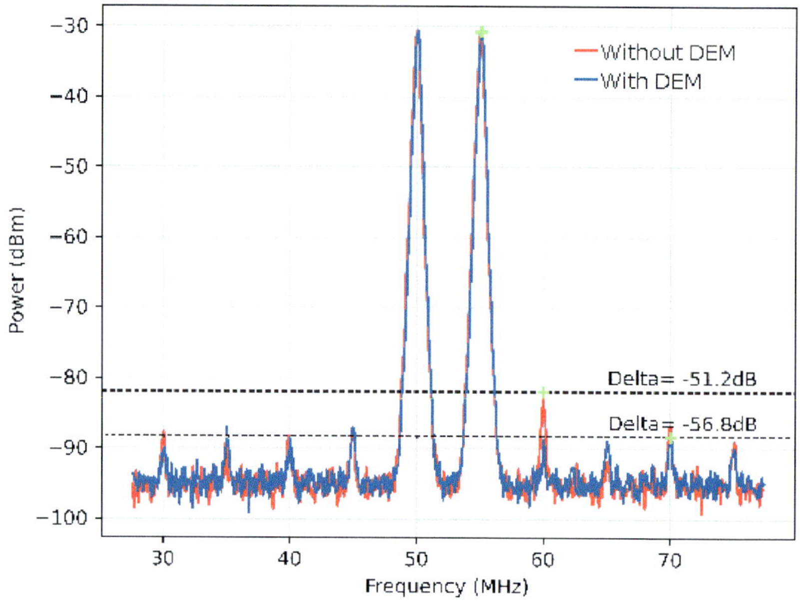

Fig. 12. Tx DAC IM3 without DEM (red) and with DEM (blue).

VII. LOCAL OSCILLATOR MULTIPLICATION

Concept

The aim of the local oscillator (LO) multiplication circuit is to provide a 39 GHz frequency source (Sx) signal needed for both quadrature modulation and demodulation in the transceiver, whilst the external reference LO signal can be distributed at a lower, subharmonic frequency.

Because LO distribution over the many channels of the full transceiver array is done off-chip and outside the RF signal band, interference from near-in-band coupling of LO distribution signals into the RF signal paths is mitigated.

In an effort to balance power dissipation efficiency in both the distribution and multiplication of the LO signal, a threefold frequency multiplication factor was chosen for this design. This is achieved by means of harmonic distortion of the input using a non-linear amplifier and various filtering and buffering stages.

The multiplier circuit converts a 13 GHz single-ended external reference signal to a 39 GHz quadrature output.

Implementation

The Sx block is implemented by means of three amplification stages, two intermediate filter stages, an input balun, two output transformers, and a splitter, shown in Figure 13.

The 13 GHz externally generated LO signal is split and distributed off-chip through the printed circuit board (PCB) and package/interposer over all transceiver channels. Each channel on the chip has an external 13GHz signal that is provided at the input pad of the circuit. The balun at the input performs single-ended to differential conversion and 50 Ohm matching, while offering ESD protection, AC coupling and DC biasing. It protects

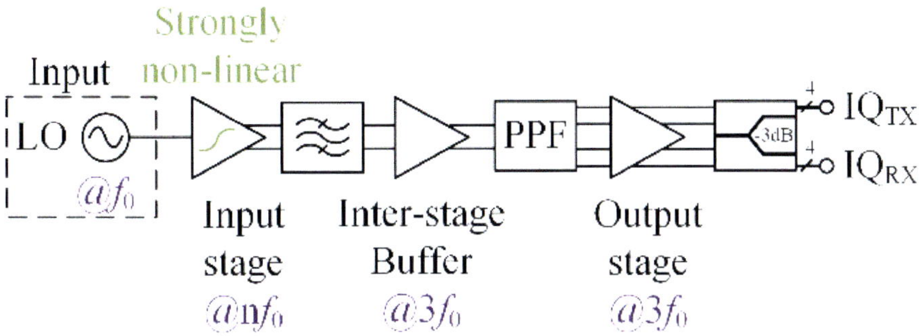

Fig. 13. Frequency multiplier block diagram.

the transistors of the input stage amplifier, and allows this amplifier to be biased independently from the source signal.

The input amplifier stage is operated in the strongly non-linear regime, and generates harmonic spurs at integer multiples of the fundamental input frequency f_0 = 13 GHz. The third harmonic has to be preserved with minimal insertion loss while the unwanted harmonic spurs have to be filtered out to avoid distortion in the later stages. This is done by ensuring a bandpass filter (BPF)-like behavior by both carefully tuning the active components, as well as matching the inter-stage transformers around the frequency of interest.

The resulting 39 GHz signal is then buffered before a poly phase filter (PPF) derives in-phase and quadrature components. Finally, the quadrature signal components are buffered once more before being split and distributed to the Rx and Tx subsystems.

The amplifier stages in the Sx block are a pseudo-differential common-source mm-wave amplifiers. The first amplifier stage is depicted in Figure 14. It is accompanied by transformers T_{1-2} at their respective inputs and outputs to provide matching, AC coupling and DC access points for supply V_{DD} and biasing V_b. Transistors M_{1-2} represent the input common-source devices that amplify the input signal, whereas transistors M_{3-4} are used as neutralization capacitors to improve the differential-mode stability. Large resistors R_l and R_b are employed to improve the common-mode stability of the ampli-

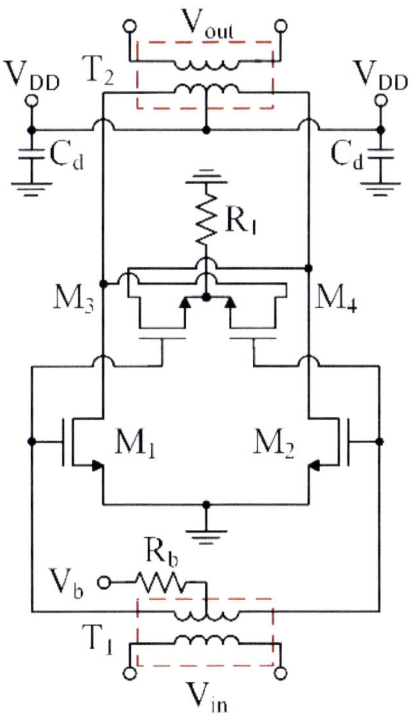

Fig. 14. Amplifier stage.

fier. Decoupling capacitors C_d are spread across unused areas of the layout, in order to filter the noise coupling from the power supply.

The same is true for the transformers and transmission-lines at the output stage, which provide output power splitting and 100 Ohm differential output matching. Moreover, the routing lengths of the quadrature components to the Tx and Rx blocks have been carefully matched to maintain phase coherence between the quadrature components at the IQ_{Tx} and IQ_{Rx} output locations.

Each DC bias point is controlled and fine-tuned by voltage digital-to-analog converters (VDACs), consisting of an 8-bit R-2R resistive ladder and allows optimization of the matching and gain properties of the amplifier stages. These voltages can be monitored externally using an analog multiplexer.

The passive components of this circuit have been designed based on the thick traces of the upper metal layers with low metal density areas to mitigate Q-factor degradation. They have been individually verified using 2.5D electromagnetic (EM) simulation. The active components are designed in accordance with noise figure (NF) guided current density optimization for the target f_{max}. The models derived from EM simulation together with the parasitic RC extraction of active components have undergone pre-silicon verification based on transient, AC (s-parameter) and harmonic balancing simulation.

In simulation and typical conditions, the Sx block shows a 53 mA current is consumption at P_{in} = –6dBm at 13 GHz from a power supply of 800 mV, with a P_{out} of 2 dBm at 39 GHz before the power splitter, resulting in a conversion gain of 8 dB. Furthermore, PVT corner simulations have shown that harmonic rejection remains above 30 dB up to the seventh.

At the time of writing, silicon validation is yet to be performed.

VIII. INPUT/OUTPUT INTERFACE

Millimeter wave, massive MIMO, and digital beamforming form a scenario where digital interconnect efficiency is critical. To enable full digital beamforming, every antenna needs to be fed with an individual data-stream and the received data from every antenna need to be recovered and processed in a central base-band processing unit. With the high number of antennas and signal bandwidth in the hundreds of MHz range, the amount of data that needs to be transferred between RF transceivers and digital base-band processing units are enormous. Energy efficiency, in terms of energy per bit, and low PCB footprint is crucial to enable big, massive MIMO arrays, with digital beamforming capabilities. In this section, we present a circuit implementation that takes on an alternative approach to digital IO that the traditional one in an attempt to create a system with a different set of design constraints, exploring the potential for improved energy efficiency with low PCB footprint and robustness (BER requirements) on a par with traditional systems.

Traditionally, digital-IO transmission is implemented using low-complexity encoding and very wideband signals. As bitrates have increased, inter symbol interference caused by channel imperfections, typically channel low-pass filter characteristics but also interface impedance mismatch caused reflections, jeopardize signal integrity. This has forced the implementation of pre-distorting and receiver equalizer implementations with increasing complexity, cost and power-consumption. To explore an alternative (partly previously explored [21][22]) approach, and explore the potential of this, we have implemented a different kind of digital transceiver that explores that, from a radio channel perspective, a PCB trace is a very good channel. Compared to a typical radio channel, the attenuation is low, blocker signals are not present, and any interference levels can be suppressed to a level far below the signal level. Thereby the available channel capacity of a PCB trace is huge. Traditional IO use "baseband" signals, with frequency content from close to DC to the effective symbol rate, this spans several octaves of frequency range. If the same type of signal is frequency translated by, for instance, mixing with a carrier frequency, the relative bandwidth (that is the ratio between carrier frequency and signal bandwidth) can be changed to a fraction of an octave. With a signal of relative narrow bandwidth, traditional radio techniques can be utilized to mitigate the effects of imperfect channel frequency characteristics and impedance mismatch effects. This is under the assumption that a PCB channel can be implemented with a flat frequency response, over a limited frequency. Whether this approach will give any power benefits in relation to the traditional wideband, low complexity symbol encoding, high complexity channel mitigation techniques, relies on the cost of implementing a high order symbol encoding and frequency translating send and receive functionality. The implemented transceiver, as shown in Figure 15, intends to explore these aspects.

Implementation

The signal energy bandwidth is made relatively narrow by moving it up in frequency by mixing. For this transceiver, we have chosen a quadrature mixer topology. This enables an easier and more efficient design supporting higher order modulation. The modulation scheme chosen is direct QAM symbol signaling. Transmission symbols are created in two 3-bit current steering DACs with up to 3.9 GS/s and 30 dB programmable gain range. The input data word is directly mapped to the binary weighted digital inputs of the two DACs. The DACs are clocked at symbol rate and is intended to be synchronous with a digital data source block (a test-pattern generator block in the test-chip). The DAC outputs are connected directly to a quadrature mixer with impedance matched output nodes. The traditional reconstruction filter is here omitted since out-of-band-emitting-requirements are not present in this transmission concept.

The LO-signal frequency is generated by an integer PLL. The PLL features a

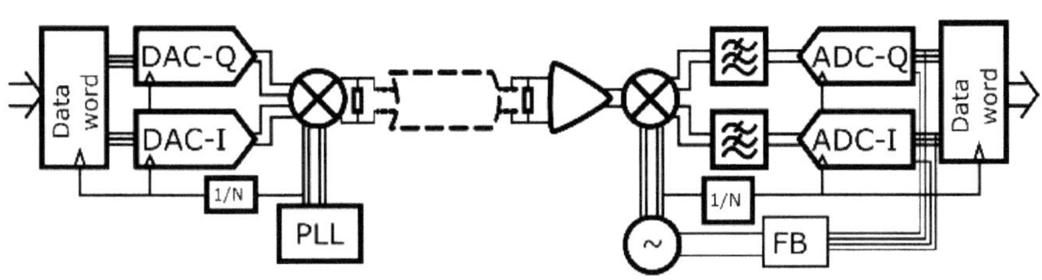

Fig. 15. IO-structure.

bang-bang phase detector and fast lock functionality using frequency phase detection and time to digital conversion detection and with delta-sigma controlled quadrature DCO oscillating at 11.7 GHz. There is an integer division factor between the LO-mixing clock and the symbol clock. This will eventually enable the tracking, and recover the symbol rate with the LO signal in the receiver.

The receiver consists of a RF gain stage and a quadrature mixer. Each baseband mixer output are connected to a passive low-pass filter followed by a flash ADC. The RF gain stage provides roughly 20 dB of fixed gain. The purpose of the filter is to suppress the sum-frequencies of the mixer. The concept of the receiver is to track the phase of the received carrier signal and rotate the LO-signals into the mixer, so that the I- and Q-baseband paths are orthogonal and correct with respect to the sent quadrature constellations. Given the correct phase of the baseband signals, the sequence of sent digital words will be presented directly at the ADC output. The ADCs are two 4-bit flash ADCs with individual programmable comparator shutdown. The ADC will enable detection of QAM4, QAM16 and QAM64 constellations as well as extra levels used for mixer phase tracking and amplitude tracking. The receiver PLL is locked when a known pilot symbol is sent by the transmitter. The outputs of the ADCs are used to detect frequency and phase errors, and through a digital feedback block control a quadrature DCO oscillating identical to the one in the transmitter. After locking, the transmitter will start transmitting data symbols and the extra detection levels in the ADCs are used to create a bang-bang feedback signal to the feedback block, keeping the receiver phase locked to the transmitter.

Expected Performance

The circuit has been verified with a channel model corresponding to microstrips in packages, two PCBs through VIAs and 300 mm impedance matched PCB trace. The model gives 12 dB of signal attenuation at the carried frequency of 11.8 GHz. With this channel, the expected power dissipation breakdown is shown in Table 3. From this state the transmitter can increase the signal gain by up to 8 dB at the expense of up to 4 mW extra power dissipation or decrease the signal gain by down to 22 dB with a decreased power dissipation of 3 mW. The ADC power scales with modulation and symbol rate in a range of [–0.8 mW to + 2 mW] from the values given in Table 3 which is at QAM16 and 2.3 GS/s. A simulated eye diagram at the ADC inputs are shown in Figure 16. Note that the I and Q signals are overlayed in the plot. Supported bit-rates and estimated energy per bit performance are shown in Tables 4 and 5 respectively. As a comparison, [23] includes a summary of recent comparable reported serial links, wherein energy per bit is reported between 2.9 pJ/bit and 5.8 pJ/bit. The energy per bit performance of this work is comparable to the reported serial links. The core circuit area is 0.1 mm^2 for the transmitter and 0.3 mm^2 for the receiver. The circuit will soon be used in the lab to validate the concept and performance in relation to various PCB channels.

TABLE 3
BLOCK POWER BREAKDOWN

Block power dissipation [mW]	Typ
Tx (DACs, mixer)	6.8
Tx PLL	13
Rx (amp, mixer, filters, ADCs)	11
Rx PLL	16.5
In total	47.3

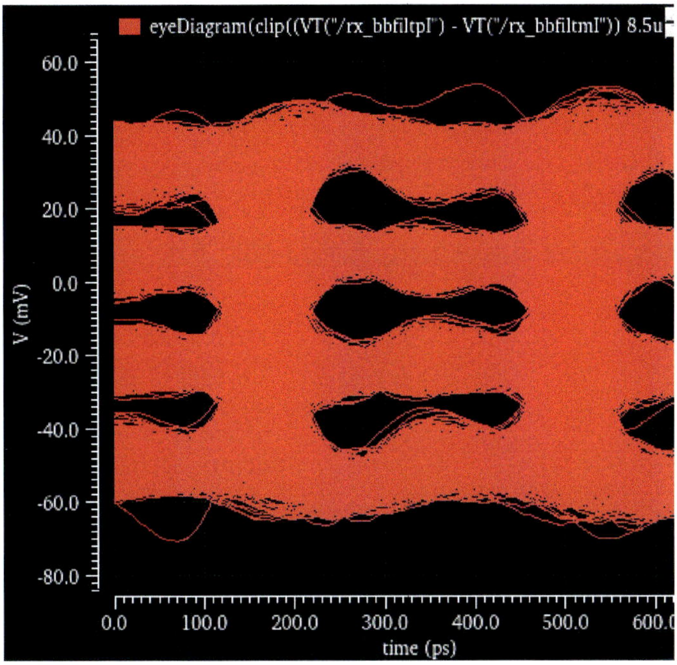

Fig. 16. Eye diagram for one of the I/Q channels receiving QAM16 @ 2.95 GS/s.

TABLE 4

BITRATES FOR SUPPORTED BITRATES AND MODULATIONS

Bitrate [Gb/s] vs. symbol rate and modulation	QAM4	QAM16	QAM64
3.93 GS/s	7.9	15.7	23.6
2.95 GS/s	5.9	11.8	17.7
2.36 GS/s	4.7	9.4	14.2
1.96 GS/s	3.9	7.9	11.8
1.47 GS/s	2.95	5.9	8.9
0.74 GS/s	1.75	2.9	4.4

TABLE 5

ESTIMATED ENERGY PER BIT FOR SUPPORTED BITRATES AND MODULATIONS

Energy per bit [pJ/bit] vs. symbol rate and modulation	QAM4	QAM16	QAM64
3.93 GS/s	5.9	3.0	2.1
2.95 GS/s	7.9	4.0	2.7
2.36 GS/s	9.9	5.0	3.4
1.96 GS/s	11.8	6.0	4.1
1.47 GS/s	15.8	8.0	5.4
0.74 GS/s	31.5	15.4	10.6

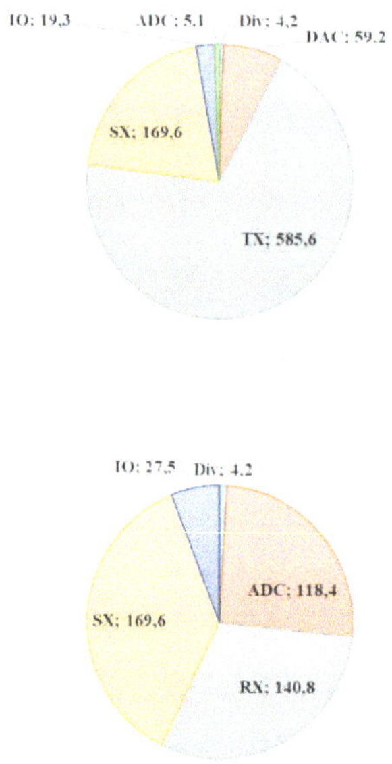

Fig. 17. Power consumption (typical) distribution per RFIC in mW (top Tx mode, bottom Rx mode). Rx mode consumption is given in regular receiving mode, not in DPD mode.

IX. CONCLUSION

The design of a complete pure digital beamforming transmitter/receiver (including digital baseband interface) has been presented. Design strategies that answer DBF challenges and take advantage of the advanced FDSOI node have been explained.

The design demonstrates the capacity of 22FDSOI to answer the requirements of dense area communication requirements just like inflight 5G communications for airplane in-cabin connectivity. Application requirements are fulfilled while keeping a moderate power consumption in both receiver and transmitter. Power consumption distribution is detailed in Figure 17.

Beyond this work some further investigation can still be achieved to address specific challenges.

- Massive MIMO leads to large antenna array, many RF ICs and huge digital baseband, that raise the issue of integration and power dissipation. A large number of RF front ends is also space consuming. This space occupation concern is not only a cost issue. Since antennas have to be separated by a λ/2 distance in the array, and since higher frequencies means lower λ, it can just be problematic to insert front ends between antennas. This has to be accounted for while choosing the number of channels per ICs. To answer that constraint an innovative approach could be to use bidirectional front ends [24]. Relaxed constraints on analog RF parts of DBF systems allows reaching the necessary trade-offs between Tx power efficiency and Rx noise figure that is inherent to bi-directional architectures.
- Regarding the need for LO synchronization, an interesting technique could be to use much lower reference frequencies. That is possible with a periodically repeated oscillations train (PROT) which also provide the advantage of reaching very low phase noise [25].
- Finally, being able to use either digital baseband outputs or IF output would allow to reach more flexibility to take even more advantage of the DBF versatility regarding system architectures. In order to limit the I/O count, polyphaser filters can be used to eliminate image frequencies inside the RFIC.

REFERENCES

[1] Al-Falahy, NFA and Alani, OYK 2019, "Millimeter wave frequency band as a candidate spectrum for 5G network architecture: a survey", *Physical Communication*, vol. 32 (Feb 19), pp. 120-144.

[2] A. N. Uwaechia and N. M. Mahyuddin, "A comprehensive survey on millimeter

wave communications for fifth-generation wireless networks: Feasibility and challenges," *IEEE Access*, vol. 8, pp. 62367-62414, 2020, doi: 10.1109/ACCESS.2020.2984204.

[3] A. Puglielli et al., "Design of energy- and cost-efficient massive MIMO arrays," *Proceedings of the IEEE*, vol. 104, no. 3, pp. 586-606, March 2016, doi: 10.1109/JPROC.2015.2492539.

[4] A. Kharalkar, A. Batabyal, R. Zele and S. Gupta, "A review of phased-array receiver architectures for 5G communications," *2022 IEEE Region 10 Symposium (TENSYMP), Mumbai, India*, 2022, pp. 1-6, doi: 10.1109/TENSYMP54529.2022.9864369.

[5] Dal Maistro et al., "A 24.2-30.5GHz quad-channel RFIC for 5G communications including built-in test equipment," *2019 IEEE Radio Frequency Integrated Circuits Symposium (RFIC), Boston, MA, USA*, 2019, pp. 283-286, doi: 10.1109/RFIC.2019.8701830.

[6] A. G. Roy et al., "A 37-40 GHz phased array front-end with dual polarization for 5G MIMO beamforming applications," *2019 IEEE Radio Frequency Integrated Circuits Symposium (RFIC), Boston, MA, USA*, 2019, pp. 251-254, doi: 10.1109/RFIC.2019.8701767.

[7] D. Dosluoglu, K.-D. Chu, D. Pena-Colaiocco, I. Zhao, V. Sathe and J. C. Rudell, "A reconfigurable digital beamforming v-band phased-array receiver," *ESSCIRC 2022- IEEE 48th European Solid State Circuits Conference (ESSCIRC), Milan, Italy*, 2022, pp. 493-496, doi: 10.1109/ESSCIRC55480.2022.9911486.

[8] S. Mondal, R. Singh, A. I. Hussein and J. Paramesh, "A 25-30 GHz 8-antenna 2-stream hybrid beamforming receiver for MIMO communication," *2017 IEEE Radio Frequency Integrated Circuits Symposium (RFIC), Honolulu, HI, USA*, 2017, pp. 112-115, doi: 10.1109/RFIC.2017.7969030.

[9] Y. Cho et al., "A 16-element phased-array CMOS transmitter with variable gain controlled linear power amplifier for 5G new radio," *2019 IEEE Radio Frequency Integrated Circuits Symposium (RFIC), Boston, MA, USA*, 2019, pp. 247-250, doi: 10.1109/RFIC.2019.8701791.

[10] M. Johnson et al., "A 4-element 28 GHz millimeter-wave MIMO array with single-wire interface using code-domain multiplexing in 65 nm CMOS," *2019 IEEE Radio Frequency Integrated Circuits Symposium (RFIC), Boston, MA, USA*, 2019, pp. 243-246, doi: 10.1109/RFIC.2019.8701732.

[11] U. Kodak, B. Rupakula, S. Zihir and G. M. Rebeiz, "A 62 GHz Tx/Rx 2x128-element dual-polarized dual-beam wafer-scale phased-array transceiver with minimal reticle-to-reticle stitching," *2019 IEEE Radio Frequency Integrated Circuits Symposium (RFIC), Boston, MA, USA*, 2019, pp. 335-338, doi: 10.1109/RFIC.2019.8701863.

[12] A. A. Alhamed and G. M. Rebeiz, "A global multi-standard/multi-band 17.1-52.4 GHz Tx phased array beamformer with 14.8 dBm OP1dB supporting 5G NR FR2 bands with multi-Gb/s 64-QAM for massive MIMO arrays," *2021 IEEE Radio Frequency Integrated Circuits Symposium (RFIC), Atlanta, GA, USA*, 2021, pp. 99-102, doi: 10.1109/RFIC51843.2021.9490458.

[13] R. Lu, C. Weston, D. Weyer, F. Buhler, D. Lambalot and M. P. Flynn, "A 16-element fully integrated 28-GHz digital RX beamforming receiver," *IEEE Journal of Solid-State Circuits*, vol. 56, no. 5, pp. 1374-1386, May 2021, doi: 10.1109/JSSC.2021.3067504.

[14] H.-C. Park et al., "4.1 A 39GHz-band CMOS 16-channel phased-array transceiver IC with a companion dual-stream IF transceiver IC for 5G NR base-station applications," *2020 IEEE International Solid-State Circuits Conference - (ISSCC), San Francisco, CA, USA*, 2020, pp. 76-78, doi: 10.1109/ISSCC19947.2020.9063006.

[15] F. Tillman, P. Paliwal, and D. Eckerbert, "A low power 5G access point targeting airplane cabin connectivity" *ESSCIRC*

2023- IEEE 49th European Solid State Circuits Conference (ESSCIRC), Lisbon, Portugal, 2023

[16] NGNM Alliance, *5G White Paper*, Feb. 2015

[17] 3GPP, "User Equipment (UE) radio transmission and reception," *Technical Specification* 38.101-2 V15.8.0 , Release 15, 2019.

[18] G. Mangraviti, B. Debaillie and P. Wambacq, "A 39-GHz 18.5-mW LNA with T/R switch, 15.4-dB gain, -2.2dBm IIP3, 5.6-dB NF, for a 5G in-cabin basestation in 22-nm FD-SOI," *2022 24th International Microwave and Radar Conference (MIKON), Gdansk, Poland*, 2022

[19] L. H. de Carvalho Ferreira and T. C. Pimenta, "A CMOS voltage reference for ultra low-voltage applications," *IEEE ICECS*, 2005.

[20] Y. Yi, D. Zhao and X. You, "A Ka-band CMOS digital-controlled phase-invariant variable gain amplifier with 4-bit tuning range and 0.5-dB resolution," *IEEE RFIC*, 2018.

[21] Y. Du et al., "A 16-Gb/s 14.7-mW tri-band cognitive serial link transmitter with forwarded clock to enable PAM-16/256-QAM and channel response detection," *IEEE Journal of Solid-State Circuits*, vol. 52, no. 4, pp. 1111-1122, April 2017, doi: 10.1109/JSSC.2016.2628049.

[22] W.-H. Cho et al., "10.2 A 38mW 40Gb/s 4-lane tri-band PAM-4 / 16-QAM transceiver in 28nm CMOS for high-speed memory interface," *2016 IEEE International Solid-State Circuits Conference (ISSCC), San Francisco, CA, USA*, 2016, pp. 184-185, doi: 10.1109/ISSCC.2016.7417968

[23] Z. Jiang, H. Beshara, J. Lam, N. Ben-Hamida and C. Plett, "High speed DMT for 224 Gb/s and faster wireline transmission," *IEEE Transactions on Circuits and Systems I: Regular Papers*, vol. 70, no. 4, pp. 1758-1771, April 2023, doi: 10.1109/TCSI.2023.3234920.

[24] L. Paquien, B. Martineau, D. Belot and N. Deltimple, "A compact bidirectional reconfigurable 2nd-order low-pass filter for 5G FR2 applications," *2023 18th Conference on Ph.D Research in Microelectronics and Electronics (PRIME), Valencia, Spain*, 2023, pp. 77-80, doi: 10.1109/PRIME58259.2023.10161920.

[25] C. Jany, A. Siligaris, J. L. Gonzalez-Jimenez, P. Vincent and P. Ferrari, "A programmable frequency multiplier-by-29 architecture for millimeter wave applications," *IEEE Journal of Solid-State Circuits*, vol. 50, no. 7, pp. 1669-1679, July 2015, doi: 10.1109/JSSC.2015.2411623.

Jérôme Prouvée received his degree in engineering and M.Sc. degree from INPG Grenoble, France in electronics and radiofrequencies in 2000.

After an internship in model extraction at STMicroelectronics Crolles France, and his military service in the RF lab of ENSTA Paris, he joined the emerging RFIC design team of CEA Leti in 2001. He spent 14 years in testing, mostly specialized in RF R&D IC characterization, test projects management and industrialization tests. He changed to custom and RF IC layout 6 years ago. He now splits his time between IC layout and project management. His main technical interest involve RF test, RF and mixed signal layout and custom/analog IC TOP assembly.

Giovanni Mangraviti received his B.Sc. and M.Sc. degrees in Electrical Engineering from the University of Messina (Università degli Studi di Messina), Italy in 2005 and 2008, respectively, and his Ph.D. degree from the Free University of Brussels (Vrije Universiteit Brussel), Belgium, in 2015.

In 2007, he was an intern with imec, Leuven, Belgium, where he was involved with the characterization and investigation of advanced low-k dielectric materials for back-end-of-line. Since 2009, he has joined the mm-Wave R&D Team of imec. Since 2019, he has been serving as TPC member of the IEEE Asian Solid-State Circuits Conference.

Investigation, design, measurement and advance of mm-wave integrated core circuits and transceivers currently involve his research activities and passion.

Björn Debaillie leads imec's collaborative R&D program on cutting-edge IoT technologies, covering high speed communications, high resolution sensing, and neuromorphic computing. As a seasoned researcher and manager, he is responsible for strategic collaborations and partnerships, innovation management, and public funding policies as well as the operational management and coordination across imec's collaborative programs and projects. Björn Debaillie coordinates public funded projects and seeds new initiatives. He holds patents, has received awards and authored books and international papers published in various journals and conference proceedings.

Piet Wambacq (Senior Member, IEEE) received his M.Sc. degree in electrical engineering and Ph.D. degree from Katholieke Universiteit Leuven, Leuven, Belgium in 1986 and 1996, respectively. In 1996, he joined imec, Leuven, where he is currently Fellow, working on analog/RF/mm-wave IC design in various technologies for wireless applications. Since 2000, he has been a Professor with Vrije Universiteit Brussel (VUB), Brussels, Belgium. He has authored or co-authored six books and more than 350 articles in edited books, international journals and conferences. He was a member of the Program Committee of ISSCC from 2012 to 2020 with the role of RF Subcommittee Chair from 2016 to 2020. He is chair of the program committee of ISSCC 2023. He was a member of the Program Committee of the ESSCIRC and DATE conferences, an Associate Editor for the IEEE Transactions On Circuits And Systems—Part 1 from 2002 to 2004, and a Distinguished Lecturer of the IEEE Solid-State Circuits Society of IEEE from 2016 to 2018.

David Borggreve received his B.Sc. and M.Sc. degrees in electrical engineering from the University of Twente, Enschede, The Netherlands in 2007 and 2010, respectively. From 2011 to 2014, he was with ST Microelectronics, Crolles, France, working as a research engineer on integrated circuit design for wireless communication applications. In 2014 he joined the Circuit Design group of Fraunhofer EMFT in Munich, Germany, where he developed mixed-signal circuits for advanced CMOS process technologies, e.g. SAR A/D Converters for a fully depleted silicon on insulator (FD-SOI) process. Since 2019 he has been a group manager of the integrated circuit design team at Fraunhofer EMFT. His research interests include mixed-signal/rf integrated circuits and neuromorphic sensing/computing.

Dr. Radu Ciocoveanu, received his B.Sc. degree in microelectronics from Technical University Gheorghe Asachi Iasi, Romania, in 2014, and his M.Sc. degree in microelectronics and microsystems from the Hamburg University of Technology (TUHH), Germany, in 2016. From 2016 to 2020, he was with Infineon Technologies AG, Munich, in cooperation with Friedrich–Alexander University Erlangen–Nuremberg, Germany, where he worked on his Ph.D., focusing on millimeter-wave (mmwave) CMOS transceiver frontend-circuits for radar and communication systems. From 2020 to 2021 he was a research scientist for the design of integrated RF circuits with Fraunhofer EMFT. Currently he is a system and applications engineer at NXP Semiconductors in Munich. His research interests include systems and mm-wave circuit design in CMOS technologies for wireless communication and radar systems.

Henrik Fredriksson received his M.Sc. in applied physics and electrical engineering in 2000 and Ph.D. in circuit design in 2008, both from Linköping University, Linköping, Sweden. Since 2000 he has worked in companies Ericsson Microelectronics, Infineon Technologies, Ericsson modem platforms, ST-Ericsson, Ericsson AB and has held since 2018 a position as master researcher Data Converter Technologies Research team, at Ericsson Lund Sweden. His research interests include systematic analog/mixed-signal IC design, TRx architectures, and high-performance data converters. He has authored/co-authored some 6 patents.

Pallavi Paliwal received her B.E. degree in electronics and communication engineering from Gujarat University, India in 2006, and her M.Tech. and Ph.D. degrees from IIT Bombay, India in 2018. From 2006 to 2008, she worked with eInfochips, India as an ASIC Design Engineer. She is presently a Researcher with Ericsson, Lund, Sweden. Her current work experience includes frequency synthesizers and transceiver systems.

Fredrik Tillman received his M.Sc. and Ph.D. degrees in Electrical Engineering from Lund University in 2000 and 2005, respectively. After graduation he joined Ericsson Mobile Platforms and participated in the first cellular modem CMOS radio development in Sweden and US before moving on to the research branch of the company. Today he heads a department at Ericsson Research with focus on integrated radio circuit design for both cellular infrastructure and device connectivity. Besides being responsible for internal R&D activities, Fredrik is active in the European research community and has been the Ericsson driver for multiple collaboration projects within the Horizon 2020 framework. He is today the technical manager for the COREnext project. Since 2022, Fredrik has served as a board member of the Swedish strategic innovation program Smarter Electronic Systems to increase competitiveness and growth in Swedish industry.

Burak Dündar (bdundar@mkr-ic.com) received his M.Sc. degree in Electronics Engineering from Istanbul Technical University with a focus of integrated circuit design in 2008. He has over 15 years of experience on analog IC design field and worked for companies like Hittite Microwave (acquired by ADI) and Dialog Semiconductor (acquired by Renesas). Since 2022 he has been working for MKR-IC. His fields of expertise include high speed phase frequency detectors, amplifiers, clock data recovery circuits and power management circuits that include DC-DC converters and linear regulators.

Hasip Terlemez (terlemezh@itu.edu.tr) is an electrical engineer specializing in integrated circuit design, with a strong focus on high-performance ADC design. He holds an M.Sc. degree in Electrical Engineering from Istanbul Technical University. With extensive expertise in data converter design, Hasip excels in analog and mixed-signal circuit design, layout and verification, employing industry-standard VLSI design methodologies and tools to work on high performance SoCs in advanced process nodes. He actively seeks new opportunities to contribute his expertise to cutting-edge projects and make a positive impact in the field.

Vincent Pinon received his Engineering degree and M.Sc. in Electronics in 2003, and Ph.D. in 2006, from the National Institute of Applied Sciences (INSA) of Lyon, France.

In 2003 he joined STMicroelectronics R&D center in Crolles, France, as researcher on high-speed DC-DC converters and polar transmitters, in BCD and advanced CMOS technologies. Starting from 2008, he worked for ST's central RF division, on low-power or wideband RF transceivers. In 2014 he moved to the smartcard division in Grenoble, France, designing contactless and NFC products.

Since 2020, he joined Asygn, to design ICs for various RF and MEMS sensors products.

Frederic Hasbani received his MSc. E.E. in 2000 from ENSERG in France. He worked then with STMicroelectronics R&D on Power Management solutions for RF and digital chips, and specifically on FD-SOI nodes since 2011. After 5 years with GN Hearing Denmark, developing medical ultra-low-power mixed-mode audio chips, he joined ASYGN in 2019 to expand the Design House activity in the Nordics. In the BEYOND5 5G RFIC project, he worked as a top-level designer and technical coordinator for the 6 contributing partners of two tape-outs in 22FDX.

Alexandre Ferret received his Engineering degree in Electronics in 2008 from the Ecole nationale supérieure d'électrotechnique, d'électronique, d'informatique, d'hydraulique et des télécommunications (INP-ENSEEIHT) in Toulouse, France.

In 2003 he joined CEA-Leti MEA laboratory (MEMS and Applied electronics) in Grenoble, France, where he worked on DC/DC converters for low voltage energy harvesting applications and in low-power/low-noise analog front end for neural applications (Brain Computer Interface).

He joined Asygn in 2011 where he took part on multiple RF and analog ASIC projects in advanced CMOS/BiCMOS technologies.

Cedric Dehos joined CEA-Leti, Grenoble, France, in 2003 after graduation from ESIEE Paris (Dipl.-Ing. and M.Sc.) in Telecommunication and Signal Processing.

Since then he has been involved in system level design of complex RF and digital base band circuits including RF behavioral modeling, architecture design and specification, system simulation and base band signal processing. His main topics of interest included ultra-wide band systems, digital compensation of RF imperfections, antenna beamforming. Since 2006, he has been involved in various mmw developments in CMOS or BICMOS technologies, including 24/79GHz Short Range Radar, 60GHz Wireless HD and WiGig. In 2011, he led Leti mmw designs and began moving the developments towards 5G small cells, promoting the use of mmw bands for 5G, and towards short-range chip-to-chip communications and contactless connectors. More recently, he proposed architecture for channel aggregation in D-band, targeting >100 Gbps wireless communication.

He has been involved in many collaborative projects funded by the European Commission, as well as in many industrial bilateral projects with transfer of technology. He is also active in the promotion of mmw wireless technologies for particle physics with CERN.

Baudouin Martineau received his M.S. degree in microelectronics from the Sciences University of Bordeaux 1, Talence, France, in 2004, and his Ph.D. degree in microwave and microtechnology from the University of Lille, Lille, France in 2008.

He did his Ph.D. work at the Institute of Electronics, Microelectronics and Nanotechnologies (IEMN), Villeneuve-d'Ascq, France, and STMicroelectronics, Crolles, France. His Ph.D. thesis was focused on the 65-nm CMOS silicon on insulator (SOI) potentialities for millimeter-wave wireless applications. In 2008, he joined the Technology R&D Department, STMicroelectronics, as a millimeter wave (mmW) and Radio Frequency (RF) Design Engineer. Since 2014, he has been an expert in RF and mmW design. Since early 2023, he has been in charge of the RFIC design lab of the CEA-Leti Research Institute, Grenoble, France. He has authored or coauthored more than 50 IEEE publications and more than 20 patents.

Abdelaziz Hamani (Member, IEEE) received his B.Sc. degree in electronics engineering, M.Sc. degree in telecommunication systems, and Ph.D. degree in telecommunications and signal processing from the Science and Technology University of Algiers, Algeria, in 2006, 2011, and 2017, respectively. During his Ph.D. studies, he developed circuits for RFID applications at IMEP-LAHC Laboratory, Grenoble, France. In 2017, he worked for six months as an Electronics Engineer at the Ecole Centrale of Lyon, France. He then joined CEA-Leti, Grenoble, where he is involved in developing RF and millimeter-wave circuits and systems-in-package. His research interests include high-frequency circuits design in advanced integrated technologies and high-speed systems.

Dominique Morche (Senior Member, IEEE) received his Diploma degree in engineering from the École Nationale Supérieure d'Electricité et de Radioelectricité de Bordeaux (ENSERB), in 1990, and Ph.D. degree in electronics from the Institut National Polytechnique de Grenoble (INPG), in 1994. His Ph.D. mainly focuses on sigma-delta ADC. From 1994 to 2001 he was employed with France Telecom as a Research Engineer. He has been involved in architecture and design of analog circuits for telecom application. He is currently working with CEA-Leti. His current research interests include the specification and design of RF architecture for the IoE and 5G systems.

Disruptive TRX Design for the D-band

Jose Luis González-Jiménez, Alexandre Siligaris, Abdelaziz Hamani, Cédric Dehos, Fabrice Chaix, Pierre Courouve, Guillaume Robe, Jean Baptiste David, Francesco Foglia Manzillo, Antonio Clemente, and Nicolas Cassiau

Abstract—**This paper presents an architecture for broadband wireless transceivers operating in the D-band with the aim of covering tens of GHz of band over the air with a limited bandwidth per channel at the baseband. The main idea is to combine several baseband channels in a multi-channel radio-frequency (RF) signal. This is in contrast with the more conventional trend of using a single channel from baseband (or intermediate frequency) to RF. In this work, the main challenges introduced by this type of broadband channel bonding architecture are analyzed and some solutions are provided. A prototype implementation in CMOS 45 nm RFSOI technology and low cost PCB technology is used to illustrate the potentialities of this disruptive approach.**

Index Terms—**CMOS, D-band, high-data rate wireless, frequency generation, 16 QAM, channel aggregation.**

This project has received funding from the ECSEL Joint Undertaking (JU) under grant agreement No 876124 for BEYOND5 project and from the European Commission (EC) under grant agreement No. 101092598 for COREnext project.

J. L. González-Jiménez, A. Siligaris, A.Hamani, C. Dehos, F. Chaix, P. Courouve, G. Robe, J-B. David, F. Foglia Manzillo, A. Clemente, N. Cassiau are with CEA-Leti and Université Grenoble-Alpes, Grenoble 38340, France (contact e-mail: joseluis.gonzalezjimenez@cea.fr).

I. INTRODUCTION

IN the race to increase the wireless data rate to satisfy the growing demand in connectivity and capacity of current and future communication systems and networks, the D-band, located between 110 and 170 GHz, is becoming a strong candidate as the first wireless frequency band able to offer throughputs that are competitive with optical fiber communications. The goal is to be able to provide a data-rate of 100 Gb/s or higher at distances up to a few hundreds of meters, which are common requirements of fronthaul/backhaul links for 5G and beyond 5G networks [1,2]. Other applications include shorter distance links for kiosk or docking fast downloading and medium distance links to replace cables in a data center or even guided links using dielectric waveguides to connect the baseband unit to the remote radio head at the mobile network or fixed wireless internet access points [3].

In this context, maximizing the spectral and energy efficiencies are key objectives. The maximum data rate that can be achieved by a wireless link is ultimately limited by the available bandwidth (BW) and the noise according to the Shannon capacity limit [4]. Figure 1 shows the maximum attainable data rate for a set of

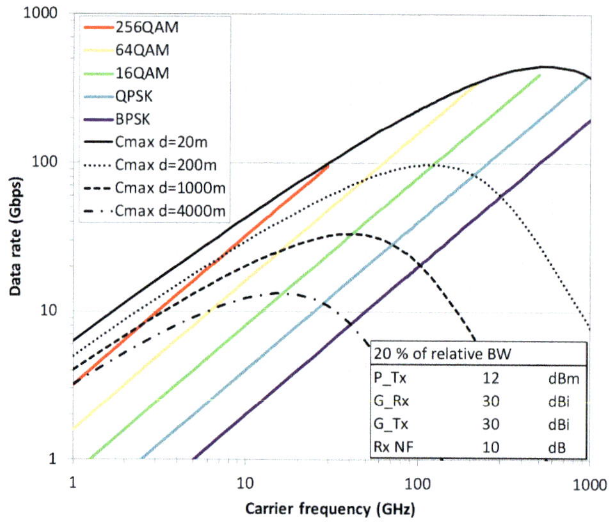

Fig. 1. Maximum link data rate as a function of carrier frequency according to the Shannon capacity limit for a 20% fractional bandwidth channel.

reasonable link parameters shown in the figure inset. In this figure, a reasonable fractional BW of 20% is considered for all carrier frequencies. The figure compares the maximum capacity of the channel for different modulation schemes. Interestingly, the peak capacity is achieved for 16-QAM at different distances and carrier frequencies. It is worth noting that 100 Gb/s can be achieved at 200 m using 16-QAM modulation. Therefore, a BW in the order of a few tens of GHz is required to achieve the 100 Gb/s data rate.

In this work, channel bonding architectures are proposed as an energy efficient alternative to conventional broadband single channel transceivers operating at the D-band. They support tens of GHz of RF BW but with much smaller baseband (BB) BW than other competing approaches. Section 2 introduces the proposed channel bonding – or channel aggregation – architecture as well as the main challenges. Section 3 presents some examples of implementation of the proposed transceiver architectures. Section 4 covers experimental point-to-point links results and the conclusions of this work and perspectives are summed up in Section 5.

II. CHANNEL BONDING TRANSCEIVERS

The basic idea behind the proposed architecture consists of building the wideband signal sent over the air by combining multiple narrower band channels. This principle is illustrated in Figure 2, where a total RF BW of 34.56 GHz is obtained by combining 16 channels of 2.16 GHz band each. The figure shows the transmitter (TX) section; the receiver (RX) is just a mirror version. Instead of bringing the 16 channels directly from BB to RF band, a two-step up-conversion scheme is preferred. It allows reducing the number of required local oscillator (LO) signals. Indeed, by grouping the BB channels in bundles of four channels and implementing a first up-conversion to intermediate frequency (IF), only four LOs are required, assuming all bundles of four BB channels share the same IF. Next, the four IF signals (containing four BB chan-

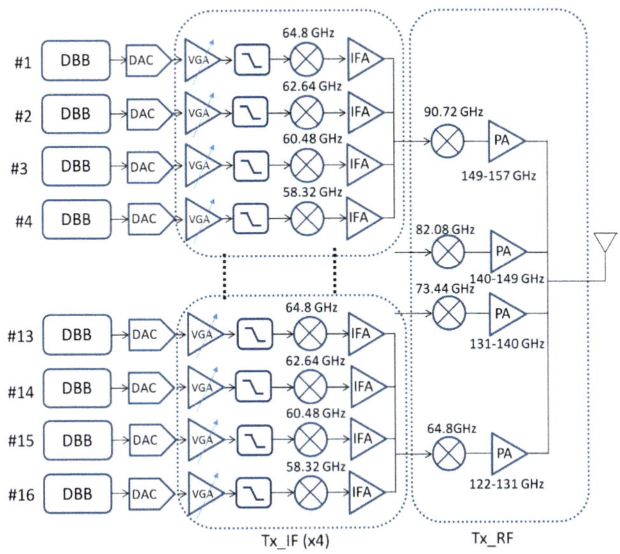

Fig. 2. Channel bonding TX architecture.

nels each) are converted to four different sub-bands at RF. This requires four additional LO frequencies. In this way, the up-conversion of the 16 BB channels only requires eight LO signals.

The careful selection of the frequency plan allows for a further property of the proposed transceiver architecture shown in Figure 2. All LO frequencies are integer multiples of the same number, 2.16 GHz, which corresponds to the channel spacing. This allows generating all LO frequencies using integer-N frequency multiplication [5] from the same common input reference. This technique has been proved to be very effective in generating multiple LO signals with adjacent channels spurs smaller than –25 dBc [6,7]. This feature is important since channel-to-channel interference in this type of transceivers should be kept below –20 dBc in order to enable a good enough signal to interference plus noise ratio (SINR) for high order modulations such as 16-QAM or 64-QAM [8]. Phase noise (PN) is also a limiting factor for high order modulations. The investigation of [8] indicates that the PN at 1 MHz offset from the carrier and the noise floor should be better than –98 dBc/Hz and –120 dBc/Hz, respectively, in order to degrade the SINR by less than 1.5 dB.

One additional challenge for this type of architecture is the way the different channel bonding operations are performed. In the proposed architecture shown in Figure 2, two channel bonding operations need to be realized. The first one happens at IF (V-band) and in the implementation of [9] it is realized using an on-chip hybrid coupler at the expenses of around 9 dB of insertion losses. The second one is done at RF (D-band) and several options are possible, such as on-chip [10], in-package [11], or over the air power combining [12]. From all these options, over-the-air power combining is the most power-efficient, as shown in [12].

III. CHANNEL BONDING TX AND RX EXAMPLES

Several implementation examples of channel bonding TX and RX modules and circuits are presented in the next subsections. They are used in the point-to-point link experiments presented in Section 4.

Baseband to IF up-converter

The first implementation example corresponds to the first up-conversion and channel aggregation step of the TX architecture in Figure 2. This circuit is used to generate each one of the IF signal four-channel bundles described in Section 2. The circuit block diagram and micro-photograph is shown in Figure 3. It constitutes four lanes with the same structure. Each lane contains an LO generator and a I/Q up-converter composed of a couple of I and Q active filters, an I/Q mixer and a voltage controlled amplifier (VGA). The output of each lane contains the corresponding BB channels up-converted to a different channel at the IF band, as shown in Figure 3. The four lane outputs are combined on-chip using a differential hybrid coupler. More circuit details can be found in [9]. The circuit is fabricated using a 45 nm RFSOI process and consumes 475 mW from a 1 V supply, occupying an area of 2 × 3.5 mm^2.

IF to D-band TX and RX modules

A first version of a two-channel TX and RX D-band circuit was implemented in 45 nm RFSOI technology. The block diagram, circuit microphotographs and module boards, including the antennas, are shown in Figure 4. The TX realizes the up-conversion and channel bonding operation for the two uppermost D-band sub-channels of the transmitter shown in Figure 2. The two IF signals are taken from the same circuit input for testing simplicity. The RX realizes the down-conversion of the same two uppermost D-band sub-channels but provides them at separate IF outputs. Both circuits are mounted on a four-layer PCB module integrating the antennas. Each of the two D-band ports of the ICs (outputs for the TX and inputs for the RX) is connected to a separate sub-array of a four-patch square antenna array implemented in the back-side of the PCB. These in-package antennas are intended to be combined with planar lenses to realize a high gain in the order of 30 dBi. More details about the antenna system can be found in [11,13] and the circuits are described in detail in [14].

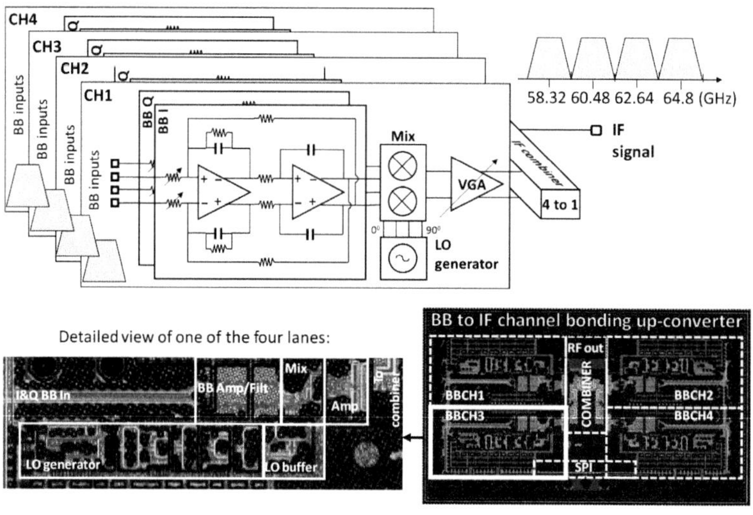

Fig. 3. BB to IF band up-converter and channel bonding circuit implementation.

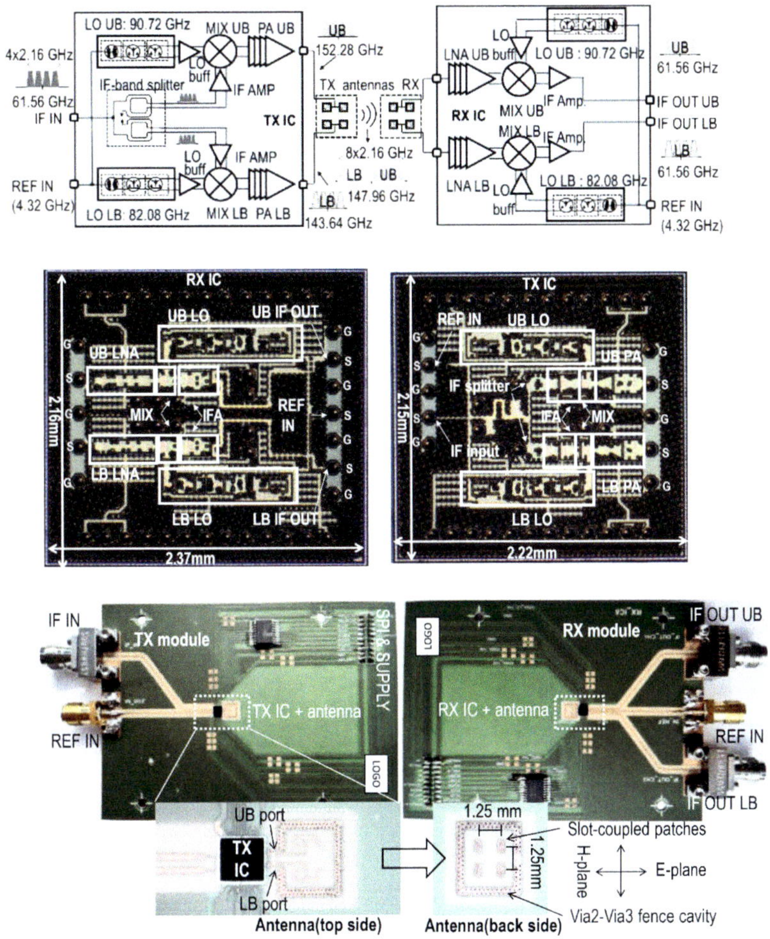

Fig. 4. D-band TX and RX modules and circuits.

The summary of TX and RX single tone measurements are shown in Figure 5 (note that the gains from IF to RF in the figure do not include the flat lens antenna gain). The effective isotropic radiated power (EIRP) at the 1 dB compression point achieved by the TX module (including the transmit array lens) is 26.4 dBm. For a multi-channel signal using high-order modulation the back-off should be larger than for single-channel signals due to a higher peak-to-average power ratio. A back-off of 9 dB has been found to be required for 16-QAM multi-channel

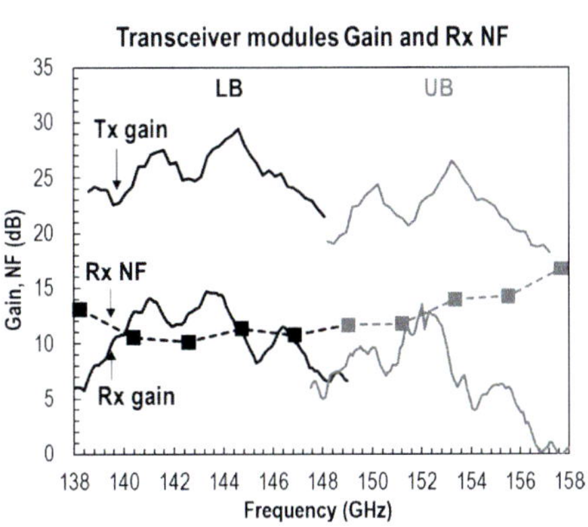

Fig. 5. D-band channel bonding TX and RX module frequency domain characteristics.

modulations in our system, resulting in an effective TX EIRP for modulated signals of 7.4 dB. The receiver module has a noise factor (NF) that varies from 11 to 15 dB across the band. The power consumption of the TX and RX circuits are 600 mW and 400 mW, respectively.

LO generation

The multiple LO signals required in the circuits presented in Sections 3.1 and 3.2 are all generated using the same technique based on integer-N frequency multiplication described in [5]. The four LO signals required for the first up-conversion operation are shown in Figure 6. The PN at 1 MHz offset is better than –100 dBc/Hz and the PN noise floor is better than –120 dBc/Hz. The LO signals for the second up-conversion step (the two highest frequency LO frequencies) are shown in Figure 7. The PN at 1 MHz offset is also smaller than –100 dBc/Hz and the PN noise floor is better than –125 dBc/Hz. Additionally, LO spurs in the adjacent channel frequencies are smaller than –30 dBc in the worst case. All these

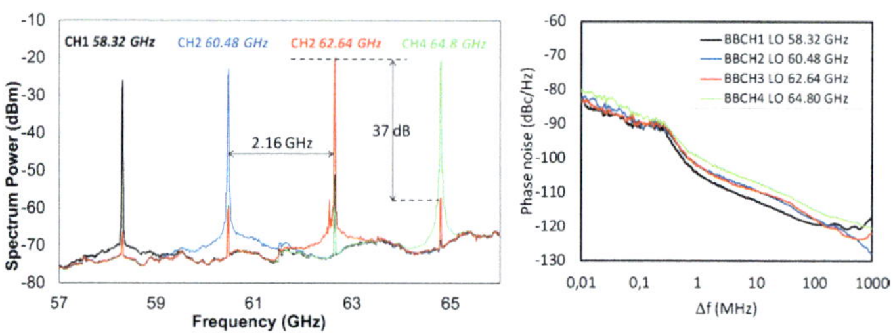

Fig. 6. LO signals spectrum and PN for the BB-IF up-converter.

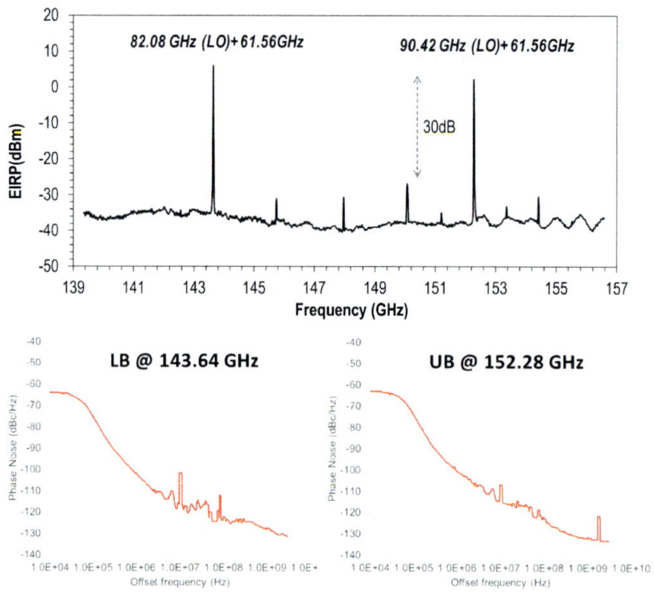

Fig. 7. LO signals spectrum and phase noise for the D-band TX and RX circuits.

features enable multi-channel operation with modulations schemes up to 64-QAM over individual channel BW of 2.16 GHz.

IV. POINT-TO-POINT LINKS EXPERIMENTS

Two point-to-point links experiments are presented in this section, the first involving the IF to D-band modules presented in Section 3.2 and the second one involving a full BB to D-band TX combining the circuits from sections 3.1 and 3.2.

Link demonstration based on D-band TX and RX modules

The TX and RX modules are inserted into a full link demonstration platform combining a MATLAB signal-processing environment that implements the digital BB (DBB) TX and RX functions and controls an arbitrary waveform generator (AWG) and a digital sampling scope (DSO) to interface with the analog components. In this first experiment, the IF signals at the input of the D-band TX module are generated by the AWG over the whole IF BW and up-converted to V-band using a commercial mixer and LO generator, as shown in Figure 8. The IF output signals of the D-band RX module are directly sampled by the DSO. The full TX-RX point-to-point link setup is shown in Figure 8. The TX module is equipped with a transmit array planar lens but not the RX module. The link distance is set to 42 cm. Figure 9 shows the signal spectrum and received constellations. The maximum data-rate is 56.32 Gb/s for 16-QAM modulation with error vector magnitude (EVM) better than 13% in all channels.

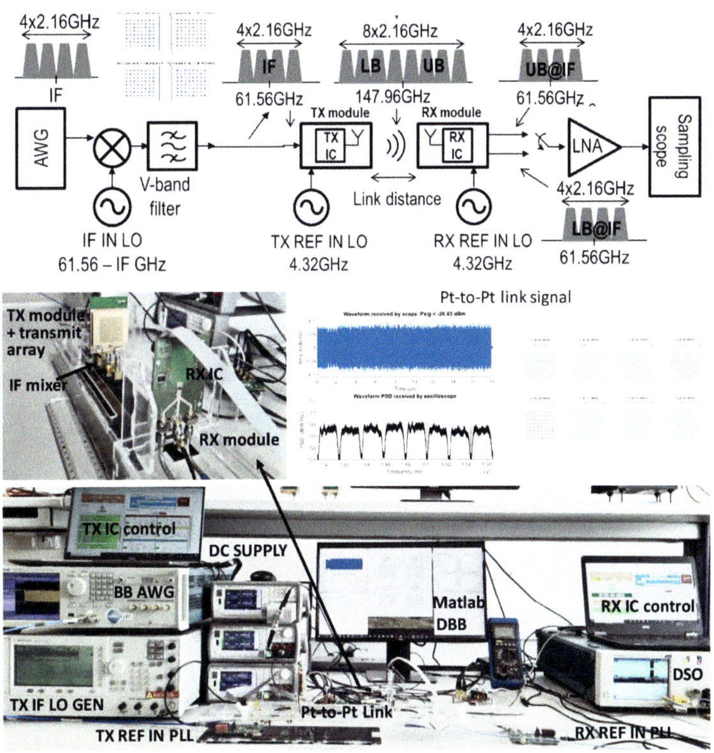

Fig. 8. Point-to-point links implemented using the D-band TX and RX modules described in Section 3.2.

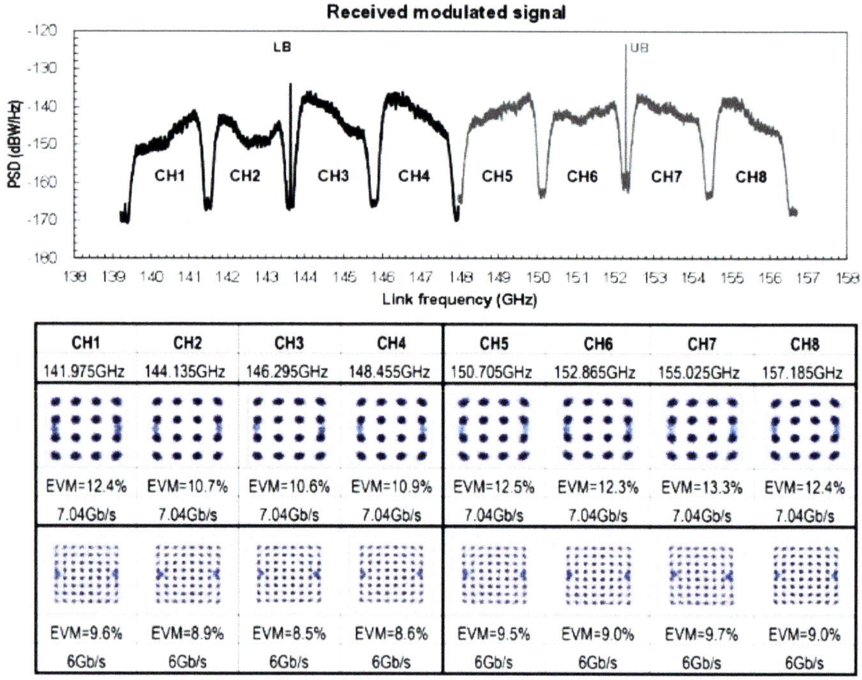

Fig. 9. D-band point-to-point link results.

Link demonstration based on BB to D-band TX

For the second experiment, the TX circuits from Sections 3.1 and 3.2 are combined in order to realize a full transmitter from multiple BB channel up to the D-band. The block diagram of the BB to D-band TX is shown in Figure 10. The RX is implemented using a commercial D-band receiver with standard 20 dBi horn antenna. The complete link set-up is shown in Figure 11. The BB signals are generated using a multi-channel AWG and a DSO is used to sample the IF signal

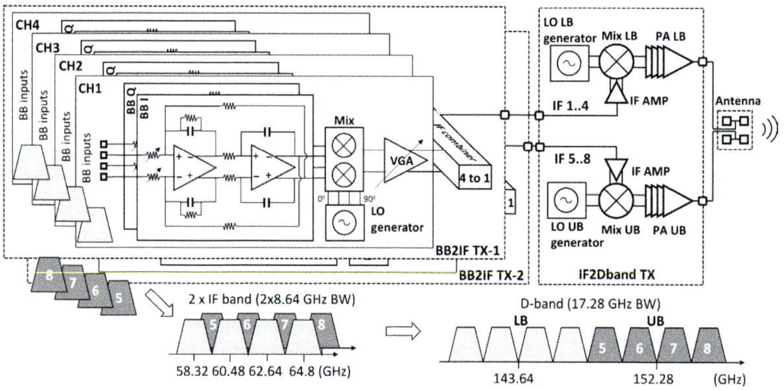

Fig. 10. BB to D-band TX block diagram.

Fig. 11. BB to D-band point-to-point link set-up.

at the output of the D-band commercial receiver. This circuit has a worst case NF of 18 dB across the signal band. Table 1 shows a link budget analysis. The SNR should be 14 dB for 16-QAM modulations in order to attain a raw bit error rate (BER) of 10^{-2} that results in an acceptable packet error rate (PER) of 10^{-4} with reasonable LDPC forward-error correction codes [15]. This is achieved at a 70 cm link distance considering the available EIRP and the RX NF numbers, with a small implementation margin.

The link shown in Figure 11 is measured for various distances. The DBB RX provides performance metrics such as signal power per channel, BER, EVM and the received constellations. The results of Figure 12 show the evolution of EVM per channel as the distance is increased from

TABLE 1
Link budget analysis

Parameter	**Per channel**	**Full band**
BW (GHz)	2.16	IF: 2 × 4 × 2.16 RF 8 × 2.16
BB I or Q diff. peak amp. (mV @ 100 Ω)	750	
BB I or Q rms amp. (mV @ 100 Ω)	270	
BB to IF up-converter gain (dB)	–29	
IF signal power (dBm)	–30.4	–24.4 (×2)
D-band TX gain (dB)	37	
EIRP@1dBCP (dBm)	17.4	26.4
Back-off (dB)	10	
EIRP for IF power (dBm)	7.4	16.4
Path Loss @ 70 cm (dB)	72.7	
Rx antenna gain (dBi)	20	
Rx NF (dB)	18	
RX signal @ antenna aperture (dBm)	–45.3	–36.3
RX noise @ antenna aperture (dBm)	–62	–53
SNR	16.7	
Margin for BER 10^{-2} on 16-QAM (dB)	2.7	

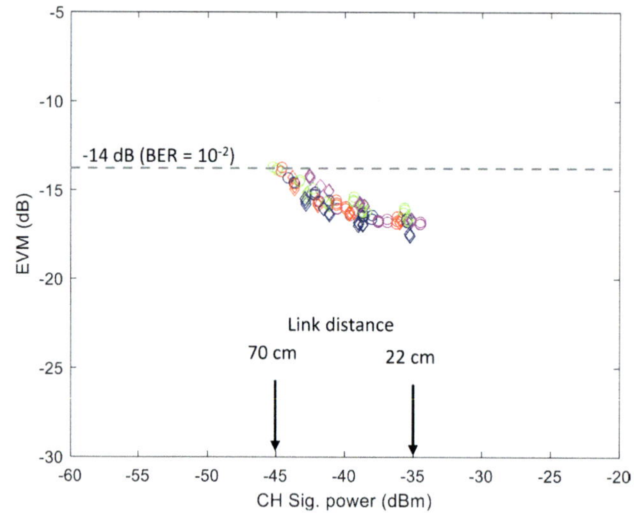

Fig. 12. EVM versus channel RX signal power for the set-up shown in Figure 11 as the TX-RX link distance is varied.

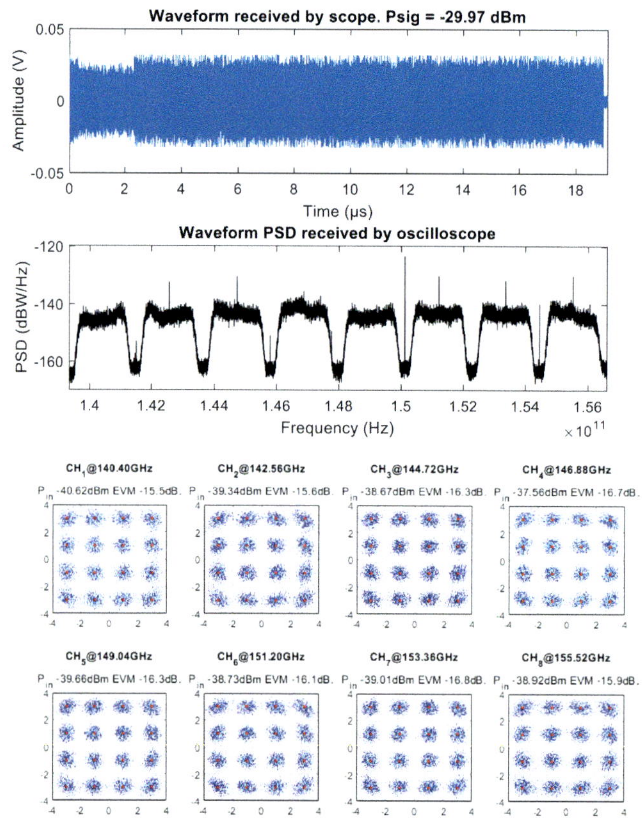

Fig. 13. RX signal waveform, spectrum and constellations for a link distance of 35 cm and 56.32 Gb/s of total data-rate.

16 to 70 cm. Note that the target EVM of –14 dB, or 20%, (for BER = 10^{-2}) is obtained for single channel power of –45 dBm, which corresponds to a link distance of 70 cm. The RX signal waveform and spectrum as well as the constellations for each channel for an intermediate distance of 35 cm are shown in Figure 13. Each channel is modulated with a 7.04 Gb/s 16-QAM signal resulting in a total link data-rate of 56.32 Gb/s and in an energy efficiency of 27.5 pJ/bit.

V. CONCLUSION

In this work, a disruptive implementation of TX and RX for high-data rate wireless communications based on channel bonding architectures is presented. The most challenging issues, including multiple LO signals generation and power combining approaches have been discussed. A full BB to D-band transceiver and two different experiments of point-to-point links have been presented showing the potentialities of these type of architectures for 100 Gb/s wireless links achieving energy efficiencies in the range of a few tens of pJ/bit.

References

[1] M. G. L. Frecassetti, *et al*, “D-Band Backhaul and Fronthaul Solutions for 5G Radio Access Network,” *2022 52nd European Microwave Conference (EuMC)*, Milan, Italy, 2022, pp. 772–775.

[2] ETSI GR mWT 012 V1.1.1 (2018-11) “5G Wireless Backhaul/X-Haul,” available on-line at: https://www.etsi.org/deliver/etsi_gr/mWT/001_099/012/01.01.01_60/gr_mWT012v010101p.pdf .

[3] Q. J. Gu, “Sub-THz/THz Interconnect, Complement to Electrical and Optical Interconnects: Addressing Fundamental Challenges Related to Communication Distances,” in *IEEE Solid-State Circuits Magazine*, vol. 12, no. 4, pp. 20-32.

[4] K. Okada, “Maximizing Energy Efficiency in Sub-THz Radio Communication and Prospective toward 6G,” *2022 IEEE BiCMOS and Compound Semiconductor Integrated Circuits and Technology Symposium (BCICTS)*, Phoenix, AZ, USA, 2022, pp. 1-4.

[5] C. Jany, *et al*., “A Programmable Frequency Multiplier-by-29 Architecture for Millimeter Wave Applications,” in *IEEE Journal of Solid-State Circuits*, vol. 50, no. 7, pp. 1669-1679, July 2015.

[6] A. Siligaris *et al*., “A Multichannel Programmable High Order Frequency Multiplier for Channel Bonding and Full Duplex Transceivers at 60 GHz Band,” *2020 IEEE Radio Frequency Integrated Circuits Symposium (RFIC),* Los Angeles, CA, USA, 2020, pp. 259-262.

[7] A. Hamani *et al*., “A D-Band Multichannel TX System-in-Package Achieving 84.48 Gb/s With 64-QAM Based on 45-nm CMOS and Low-Cost PCB Technology,” in *IEEE Transactions on Microwave Theory and Techniques,* vol. 70, no. 7, pp. 3385-3395, July 2022.

[8] J.L. Gonzalez-Jimenez, C. Dehos, N. Cassiau, *et al*. “Channel-bonding CMOS transceiver for 100 Gbps wireless point-to-point links,” in J Wireless Com Network , vol. 117, pp. 1-21, June 2020.

[9] A. Siligaris, P. Courouve, G. Waltener, A. Hamani, C. Dehos and J. L. Gonzalez-Jimenez, “A 42.24 Gb/s Channel Bonding Up-Converter with integrated multi-LO generation in 45nm CMOS,” *2023 IEEE 23rd Topical Meeting on Silicon Monolithic Integrated Circuits in RF Systems*, Las Vegas, NV, USA, 2023, pp. 4-7.

[10] A. Hamani et al., “A 84.48-Gb/s 64-QAM CMOS D-Band Channel-Bonding Tx Front-End With Integrated Multi-LO Frequency Generation,” in *IEEE Solid-State Circuits Letters*, vol. 3, pp. 346-349, 2020

[11] J.L. Gonzalez-Jimenez, *et al*., “A D-Band High-Gain Antenna Mod-

ule Combining an In-Package Active Feed and a Flat Discrete Lens," *2022 52nd European Microwave Conference (EuMC)*, Milan, Italy, 2022, pp. 784-787.

[12] J.L. Gonzalez-Jimenez, *et al.*, "A 57.6Gb/s Wireless Link Based on 26.4dBm EIRP D-Band Transmitter Module and a Channel Bonding Chipset on CMOS 45nm," *2023 IEEE Radio Frequency Integrated Circuits Symposium (RFIC)*, San Diego, CA, USA, 2023, pp. 1-4.

[13] F. Foglia Manzillo, et al., "A 28dBm-EIRP Low-Profile D-Band Transmitting Module with a Folded Transmitarray Antenna," *2023 IEEE/MTT-S International Microwave Symposium (IMS)*, San Diego, CA, USA, 2023, pp. 1-4.

[14] A. Hamani *et al.*, "A 56.32 Gb/s 16-QAM D-band Wireless Link using RX-TX Systems- in-Package with Integrated Multi-LO Generators in 45nm RFSOI," *2022 IEEE Radio Frequency Integrated Circuits Symposium (RFIC)*, Denver, CO, USA, 2022, pp. 75-78.

[15] J. Rosseel, V. Mannoni, V. Savin and I. Fijalkow, "Error Structure Aware Parallel BP-RNN Decoders for Short LDPC Codes," *2021 11th International Symposium on Topics in Coding (ISTC)*, Montreal, QC, Canada, 2021, pp. 1-5.

Jose Luis González-Jiménez (M'99, SM'12) received a Diploma in telecommunication engineering from Ramon Llull University, Barcelona, Spain, in 1992, and both an M.Sc. degree in telecommunications engineering and a Ph.D. degree (with Hons.) in electronic engineering from the Universitat Politecnica de Catalunya (UPC), Barcelona, in 1994 and 1998, respectively. He received the "Habilitation à Diriger des Recherches" degree from Grenoble University, France, in 2013. He is currently a Research Director at CEA-Leti, Grenoble, France, and invited lecturer at Phelma Engineering School, Grenoble-Alpes University. Up to 2011, he was a full-time associate professor at the Department of Electronic Engineering, Univ. Politecnica of Catalonia (UPC). From January to July 1999, he was awarded a Fulbright Fellowship and worked as a Guest Researcher with the University of Arizona, Tucson, where he collaborated with Motorola in the design of high-speed digital-to-analog converters. From September 2007 to August 2008, he was a Visiting Researcher with the STMicroelectronics/CEA-Leti joint Advanced Research Laboratory at Minatec, Grenoble, France, where he was engaged in research on the design of millimeter-wave oscillators and substrate noise coupling in millimeter-wave CMOS integrated circuits (ICs). In 2013 he was appointed as Senior Expert at CEA-LETI and since 2015 he has served as the deputy head of the RFIC design laboratory. He has led and participated in several international projects both collaborative research project and contracts with companies. He regularly collaborates as a reviewer in the journals IEEE Transactions of Circuits and Systems (I & II), IEEE Microwave and Wireless Components Letters, IEE Electronics Letters, IEEE Transactions on Microwave Theory and Techniques, among others, and serves regularly on the technical program committee of several international conferences, including IEEE IMS, ESSCIRC and NEWCAS. He is the author of two books, a book chapter, 38 international journal papers, and more than 90 conference papers. He holds 18 patents. His research interests include very large-scale integration design, mixed-signal/RF and mmW ICs, silicon photonics, and signal and power integrity in SoC and RFICs.

Alexandre Siligaris was born in Greece, in 1977. He received his M.Sc. and Ph.D. degrees from the Science and Technology University of Lille, Villeneuve-d'Ascq, France, in 2001 and 2004, respectively. During his Ph.D. studies, he was involved in transistor characterization and modeling for high-frequency applications. From 2004 to 2005, he was an associate professor with the Science and Technology University of Lille. In 2006, he joined CEA-Leti, Grenoble, France. His work focuses on high-frequency modeling and high-frequency circuit design in CMOS and BiCMOS advanced technologies.

Abdelaziz Hamani (Member, IEEE) received his B.Sc. degree in electronics engineering, M.Sc. degree in telecommunication systems, and Ph.D. degree in telecommunications and signal processing from the Science and Technology University of Algiers, Alger Ctre, Algeria, in 2006, 2011, and 2017, respectively. During his Ph.D. studies, he developed circuits for RFID applications at IMEP-LAHC Laboratory, Grenoble, France. In 2017, he worked for six months as an electronics engineer at the Ecole Centrale of Lyon, Écully, France. He then joined CEA-Leti, Grenoble, where he is involved in developing RF and millimeter-wave circuits and systems-in-package. His research interests include high-frequency circuit design in advanced integrated technologies and high-speed systems.

Cédric Dehos joined CEA Leti, Grenoble, France, in 2003 after graduation from ESIEE Paris (Dipl.-Ing. and M.Sc.) in Telecommunication and Signal Processing. Since then he has been involved in system level design of complex RF and digital base band circuits including RF behavioral modeling, architecture design and specification, system simulation and base band signal processing. His main topics of interest included ultra-wide band systems, digital compensation of RF imperfections, and antenna beamforming. Since 2006, he has been involved in various mm-w developments in CMOS or BICMOS technologies, including 24/79GHz Short Range Radar, 60GHz Wireless HD and WiGig. In 2011, he led Leti mm-w designs and began moving the developments towards 5G small cells, promoting the use of mm-w bands for 5G, and towards short-range chip-to-chip communications and contactless connectors. More recently, he proposed an architecture for channel aggregation in the D-band, targeting >100 Gbps wireless communication. He has been involved in many collaborative projects funded by the European Commission, as well as in many industrial bilateral projects with transfer of technology. He is also active in the promotion of mm-w wireless technologies for particle physics with CERN.

Fabrice Chaix was born in France in 1974. He received his M.Sc. degree in microelectronics from the Science University of Grenoble, Grenoble, France, in 2000. In 2001, he joined CIRIEL, Grenoble, France, where he worked on various research projects as an Analog Circuit Designer. In 2004, he joined EASII-IC, Grenoble, France, where he was involved with high-speed ADC circuits in designing CMOS advanced technologies. In 2008, he joined CEA-Leti, Grenoble, France, where he is involved with analog and millimeter-wave circuit design in CMOS advanced technologies.

Pierre Courouve received his M.Sc. degree from the Ecole Supérieure d'Ingénieurs de Nice Sophia-Antipolis, Nice France in 2005. He was consultant with ST Microelectronics for 7 years as Analog IC Design engineer, working on high-speed op-amp for standard purposes or rad-hard applications on SOI bipolar technology (1 or 0.25 μm). He joined Maxim Integrated, Gières, France, for 2 years as MTS Analog Engineer involved in many IPs in Power Management IC sub-systems for major internationals suppliers. Since 2014, he has been with CEA-Leti, Grenoble, France, and works in RF Architecture and IC Design Lab on RF-ID solutions and technology transfer for industrials partners thanks to his power management expertise. He is the author of 2 patents on new architecture improving data rate up to 27.12 Mb/s on HF passive tags. He skills-up in RF field through low power wide area transceiver projects and on IoT topics designing ASICs in either bulk or advanced FD-SOI technology stack. He focuses his research on narrow band filters based on N-path architectures suitable on such integrated solutions. Recently, he has followed up on industrial transfer for a narrow band transceiver for satellite LPWAN communications and UWB SoC for localization.

Guillaume Robe received his M.Sc. degree from Institut Polytechnique de Grenoble, Grenoble France in 2020. He worked for 2 years as a consultant for the CEA-Leti as an Analog IC designer, working on the power management of an UWB SoC. In 2022 he joined the CEA-Leti and works in the RF Architecture and Design Lab. His works focus on power management and analog circuit design.

Jean Baptiste David was born in 1973 and received his M.Sc. in 1997. After various experiences in industrial companies and startups, he joined CEA-Leti in 2002. He worked on passive components modelling and passive filters design on BAW and SAW technologies. He is also involved in RF-mmW passive component modeling and electromagnetic simulation. His research interest includes mm-w circuit design for frequency synthesis, including VCOs, ILOs among others.

Francesco Foglia Manzillo (Member, IEEE) received his M.Sc. degree (cum) in electronics engineering from the University of Naples Federico II, Naples, Italy, in 2012, and his Ph.D. degree in signal processing and telecommunications from the University of Rennes 1, Rennes, France, in 2017. He spent two research stays at Delft University of Technology, The Netherlands, in 2012, at the University of Michigan, USA in 2016. Since July 2017, he has been with CEA-Leti, Grenoble, France. His research interests include the analysis, synthesis, and design of antenna arrays, quasi-periodic structures, beamforming systems, and the integration of millimeter-wave radio systems. Dr. Foglia Manzillo was a co-recipient of the Best Innovation Award at the 39th ESA Antenna Workshop in 2018 and co-authored a paper awarded with the EuMC Young Engineer Prize at EuMW 2022.

Antonio Clemente (Senior Member, IEEE) received his B.Sc. and M.Sc. degrees in telecommunication engineering and remote sensing systems from the University of Siena, Siena, Italy, in 2006 and 2009, respectively, and his Ph.D. degree in signal processing and telecommunications and the "Habilitation à Diriger des Recherches" degree from the University of Rennes 1, Rennes, France, in 2012 and 2021, respectively. His Ph.D. project was realized at the CEA-Leti, Grenoble, France. From October 2008 to May 2009, he realized his master thesis project at the Technical University of Denmark (DTU), Lyngby, Denmark, where he was involved in spherical near-field antenna measurements. In 2012, he joined the Research and Development Laboratory, Satimo Industries, Villebon-sur-Yvette, France. From 2016 to 2018, he was the Technical Coordinator of the H2020 joint Europe and South Korea 5G-CHAMPION Project. Since 2013, he has been a Research Scientist with CEA-Leti. He has authored or coauthored more than 137 articles in international journals and conferences and received 20 patents. He has been involved in more than 28 research projects at the national and European levels. His current research interests include fixed-beam and electronically reconfigurable transmit array antennas, millimeter-wave, and sub-terahertz antenna-in-package (AiP), antenna arrays, periodic or quasiperiodic structures, near-field focused systems, antenna theory and fundamental limitations, synthesis and modeling, and near- and far-field antenna measurements. Dr. Clemente received the Young Scientist Award (First Prize) during the 15th International Symposium of Antenna Technology and Applied Electromagnetics (ANTEM 2012) and the Best Antenna Design and Applications Paper Award during the 13th European Conference on Antennas and Propagation (EuCAP 2109). He was a co-recipient of the EuMC Young Engineer Prize at EuMC 2021, the Best Paper Award at JNM 2015 (19emes Journées Nationales Microondes), and the 2019 ETRI Journal Best Paper Award. In 2019, he was a Finalist for the "Microwave Prize" at the European Microwave Conference (EuMC 2019). He is currently an Associate Editor of the journal Frontiers in Communications and Networks, and serves as a reviewer for the numerous IEEE and IET journals in the field of microwave, antennas, and propagation.

Nicolas Cassiau received his M.Sc. degree in Signal Processing in 2001 from Polytech'Nantes, France. Since then he has been a research engineer and project manager at CEA-Leti in Grenoble, France. His fields of interest are digital wireless communications and algorithms design. He is, or has been, working in particular on physical layer design and assessment for 4G and 5G, RF impairments modeling and sub-terahertz communications. He has been involved in several European projects, including mmMAGIC (millimeter wave communications), 5GNOW (waveform design) and more recently 5G-ALLSTAR. He has authored or co-authored over 35 papers and holds several patents in the above-mentioned fields.

Packaging Technologies and Challenges Towards 5G Integration of mm-wave Components and Silicon ICs

Tekfouy Lim, Michael Kaiser, Mattis Obst, Tanja Braun, Marius van Dijk, Johannes Jaeschke, Lars Böttcher, Kavin Senthil Murugesan, Uwe Maass, and Ivan Ndip

Abstract—**A critical part of 5G communication systems design is the packaging technology, especially of front-end components, which has to support the high-frequency and high-power requirements. This technology plays a crucial role in enabling the miniaturization, integration, and reliability of the components. In this context, this article aims to provide a workflow of the packaging technology design for 5G communication systems within the BEYOND5 project, including its key components, integration challenges, and electromagnetic simulations. The design process of an interposer begins with the selection of material, which is determined by the tolerable losses at high frequencies. After choosing the material, the bump pitch on the front-end chip and the routing complexity on the interposer determines which stack-up will be used. Thermal management is also an important step in the design process, necessary for reliable power dissipation. Furthermore, RF simulations on the selected material are to evaluate the RF performance capabilities of the interposer. As a result, the critical signal traces on the interposer are well matched at 50 Ω and arranged in order to offer low loss and reduced crosstalk (below –15 dB).**

Index Terms—**5G, base station, packaging, RF design, flip chip assembly, thermal management.**

The research leading to the results has been obtained in the scope of BEYOND5 project; it has received funding from (1) the H2020 Framework of the European Union under Grant Agreement n° 876124 and (2) the Bundesministerium für Bildung und Forschung (BMBF) with National Grant n° 16MEE0047.

All authors are with Fraunhofer IZM (Institute for Reliability and Microintegration), Berlin, Germany (e-mail: {tekfouy.lim, michael.kaiser, mattis.obst, tanja.braun, marius.van.dijk, johannes.jaeschke, lars.boettcher, kavin.senthil.murugesan, uwe.maass, ivan.ndip}@izm.fraunhofer.de). K. S. Murugusan is also with the Technical University of Berlin, Germany (e-mail: murugesan@tu-berlin.de).

I. INTRODUCTION

THE evolution of wireless communication technology has witnessed significant advancements over the years, and two groundbreaking generations that will revolutionize the way we connect and communicate are 5G and 6G. These next-generation wireless networks, conceived for their incredible speed, ultra-low latency, and high capacity, are poised to transform various industries and unlock a wide array of innovative applications.

The rapid evolution of wireless technology has opened up new possibilities for connectivity, and one such exciting application is the use of 5G technology for in-flight communication. As air travel becomes increasingly prevalent and interconnected, the need for reliable and high-speed connectivity during flights has become paramount. With its remarkable capabilities, 5G is poised to revolutionize in-flight communication, offering passengers and airlines an unprecedented level of connectivity and services [1].

Beyond passenger connectivity, 5G technology for in-flight communication also benefits airlines themselves. It enables efficient data transmission between aircraft and ground systems, facilitating real-time monitoring, maintenance, and diagnostics. This connectivity can enhance flight operations, including aircraft tracking, weather monitoring, fuel optimization, and overall fleet management. With reliable and fast in-flight communication, airlines can improve safety, enhance operational efficiency, and provide a more enjoyable travel experience for passengers.

II. SINGLE BLOCK RFIC

Within the BEYOND5 project, a demonstration of a 39 GHz millimeter wave (mm-wave) MIMO radio solution with a 50 MHz channel bandwidth is planned for the in-cabin 5G base station. This base station will offer coverage to user equipment (UE) within a range of 1–10 m. To achieve this coverage, the system will employ digital beamforming techniques utilizing a 4 × 4 MIMO array.

Demonstrator Specifications

The structure of the demonstrator, as depicted in Figure 1, comprises two main components: the active antenna system (AAS), containing the antenna array and the RF frontend with a high-speed interface to the transmitter and receiver blocks, and the digital baseband (DBB) subsystem implemented on an FPGA prototype. The AAS system is designed as a 4 × 4 array supporting dual polarization, with each array element consisting of two orthogonal ports. In total, the AAS is connected to eight RF frontend modules, where each module incorporates four RF transceivers on a single RFIC.

To meet the demanding 5G inflight broadband connectivity (IFBC) requirements, which include a download speed of 1.2 Gbps and an upload speed of 0.6 Gbps, this architecture utilizes a 64QAM modulation scheme while employing the minimum bandwidth for the n260 channel. The system utilizes a 4 × 4 MIMO array to enhance performance.

To establish communication between the RF frontend and the baseband system, a high-speed link is employed. In this setup, each of the eight RFICs is connected to the baseband system via a dedicated high-speed link. The baseband system is implemented on an FPGA platform.

For the downlink and uplink transfer, the system utilizes a 7-bit digital-to-analog converter (DAC) operating at a sampling rate of 600 MSps for downlink data transmission, and a 10-bit analog-to-digital

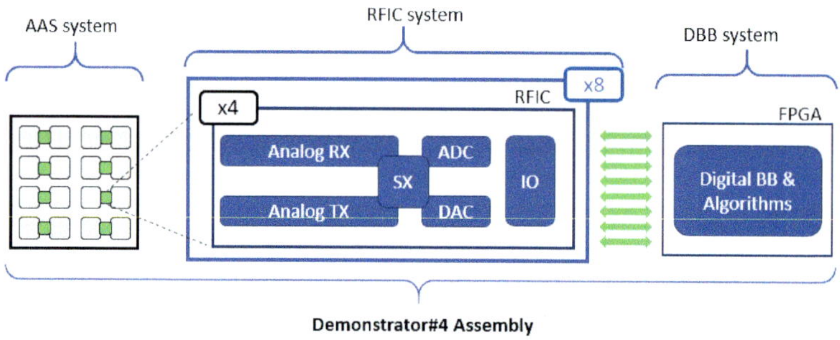

Fig. 1. System diagram of 39GHz transceiver frontend.

converter (ADC) operating at a sampling rate of 240 MSps for uplink data reception. Each of the eight high-speed links between the RFICs and the baseband system requires a data rate of 33.6 Gbps for downlink transfer and 19.2 Gbps for uplink transfer.

In the future, as the demand for higher data rates in the backhaul capacity increases, there is potential for the in-cabin 5G base station system to be designed with wider channel bandwidth, larger antenna arrays, and higher modulation schemes. These advancements would enable even faster data transfer rates and improved overall system performance.

Interposer Specifications

This paper primarily focuses on developing mm-wave modules and scalable systems by targeting an antenna-on-package (AoP) approach [2][3][4]. This approach is considered to offer the optimal balance between performance, reliability, miniaturization, and cost in RFIC packaging [5][6].

To develop the packaging concept, a collaborative effort is undertaken with RFIC designers, considering the specific requirements and properties. The packaging must not only provide a connection for the RFIC but also accommodate an antenna array positioned on top of the package. Furthermore, it needs to incorporate various signal and power distribution networks, as well as establish a defined interface with a system board.

To fulfill these requirements, extensive research is conducted on materials commonly used in printed circuit board (PCB) technology. The aim is to identify materials that possess favorable high-frequency properties. These materials play a crucial role in ensuring efficient signal transmission, reducing losses, and maintaining the desired performance levels in the mm-wave range.

By investigating and selecting suitable PCB materials with excellent high-frequency characteristics, it aims to achieve optimal performance, reliability, and miniaturization in the RFIC packaging. This comprehensive approach considers the specific needs of mm-wave modules, considering the challenges and opportunities associated with integrating the antenna, RFIC, signal/power distribution networks, and system board interface into a compact and efficient package design.

Several packaging concepts were devised and evaluated to facilitate the seamless integration of RFICs and antennas. These concepts were primarily based on the utilization of multilayer PCB technology, incorporating high-frequency laminates with minimal signal loss.

Moreover, the package design incorporates provisions for antennas positioned on the top surface. This configuration allows for enhanced antenna performance while maintaining a compact form factor.

Figure 2 showcases the packaging configuration that was selected based on comprehensive evaluation and comparison. This particular design demonstrates the optimal integration of RFICs and antennas, taking advantage of multilayer PCB technology with low-loss high-frequency laminates.

The selected packaging configuration serves as a foundation for achieving efficient signal transfer, robust power distribution, and superior antenna performance. By employing advanced packaging concepts and leveraging high-frequency PCB technology, the integration of RFICs and antennas can be realized with high performance and reliability.

This design offers a flip chip mounting approach on the package, providing several benefits. The package itself is then mounted onto a system board using a ball grid array (BGA) interface. This configuration ensures a minimal signal path with

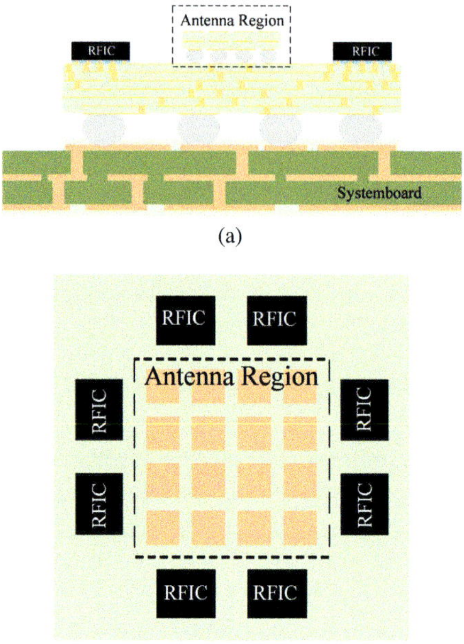

(a)

(b)

Fig. 2. BEYOND5 packaging concept. (a) Cross section, (b) top view.

few discontinuities, resulting in improved signal integrity and reduced signal loss between the package and the antenna [7][8].

One notable advantage of this design is the simplified assembly process, as the assembly steps are primarily focused on the top side of the package. This streamlines the manufacturing process, reducing complexity and potential assembly errors.

Furthermore, the accessibility of the chip from the top side offers several advantages. It allows for the convenient attachment of heat sinks or other cooling mechanisms, effectively dissipating any excess heat generated by the chip.

Additionally, the use of a separate antenna that interfaces through the BGA interface provides added flexibility. This arrangement allows for independent testing and replacement of the antenna if needed, without requiring extensive rework or modification of the package or system board.

To ensure RF compatibility and optimal performance, all interconnects within the package were meticulously designed. This involved employing full-wave electromagnetic (EM) simulations to model and optimize the interconnects. By leveraging these simulations, potential signal integrity issues, impedance mismatches, and other RF-related concerns were addressed, resulting in an optimized design that maximizes RF performance and minimizes signal degradation.

In summary, this package design offers a top-side flip chip mounting approach, simplified assembly steps, accessibility for heat sink attachment, flexibility through separate antenna integration, and rigorous RF compatibility optimization. These features collectively contribute to enhanced signal integrity, improved performance, and ease of assembly and maintenance. The package design not only considers RF compatibility but also addresses thermal management concerns. As the module operates, power losses within the ICs generate heat locally. It is crucial to implement effective thermal management strategies to ensure reliable operation.

Thermal Study of Different Packaging Concepts

Within the project, various novel packaging concepts have been subjected to numerical analysis and comparison to evaluate the thermal concept. The focus lies on optimizing the heat flow from the chip to the backside of the system board, enabling efficient heat dissipation. To facilitate the analysis and evaluation process, simplified geometries have been developed and assessed. These simplified geometries allow a quick comparison of different approaches and application scenarios, assess the impact of various factors, and derive optimization strategies.

Figure 3 illustrates several potential design concepts along with the corresponding calculated temperature rise relative to the ambient temperature, to aid the decision-making process. Symmetry was used, so only a quarter of the model is shown/simulated. The goal is to dissipate the heat generated by the power loss from the dies located on the topside of the interposer to the bottom of the system board. As the first simulation result indicates, sophisticated thermal management is needed to reduce the maximum occurring temperature. Three different thermal management concepts were analyzed. The concept with the soldered metal clips already brings a significant reduction in occurring temperature. For this variant, the clips are soldered on both the dies and system board, having thermal vias below the soldered interconnection. The variant having metal clips mounted through the system board behaves better. Between clip and die, a pressure contact is present, which has mechanical and processing limitations. The final concept, where copper ribbon-bonds directly connect the dies with copper islands in the system board, shows the best behavior.

The technological feasibility of these design concepts is evaluated in subsequent steps. The objective is to determine the viability and practicality of implementing these designs, considering factors such as manufacturing capabilities, material availability, and overall system requirements.

By considering both RF and thermal aspects in the package design, the project aims to achieve optimal performance, reliability, and thermal stability. The ongoing analysis and evaluation processes will lead to refined designs and improved thermal management strategies, ultimately ensuring the module's efficient operation under varying operating conditions.

III. STANDALONE BLOCK RFIC

Before designing a complete transceiver, an RFIC is developed with standalone single blocks. The RF integrated circuits can be measured on a wafer prober with RF probes. However, certain circuits need

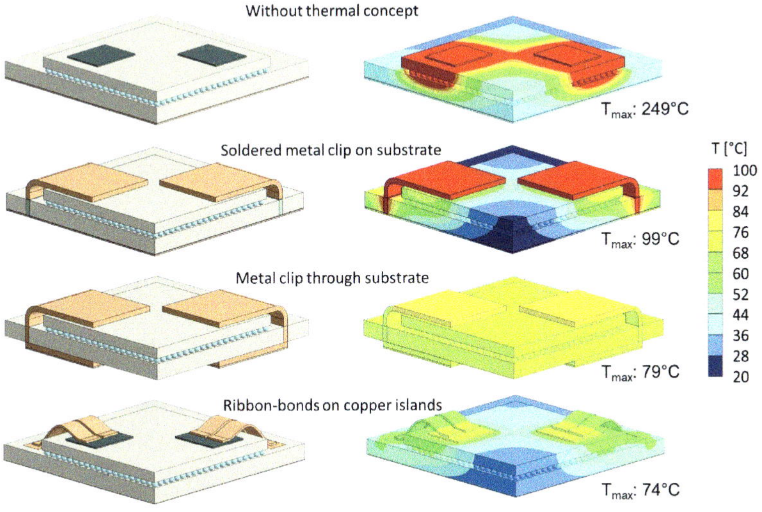

Fig. 3. Possible setup concepts with the corresponding calculated temperature rise.

to be connectorized for the characterizations and the single block RFIC needs to be fanned out via an interposer. The main focus in this section is the design of interposer and the assembly of the RFIC and various decoupling capacitors on the interposer. The development of this interposer and its assembly process will help to minimize the risks on the development of the demonstrator interposer.

The single block RFIC comprises multiple blocks sourced from different partners, including an ADC from MRK-IC, a DAC from Asygn, and an IO from Ericsson. To facilitate the characterization of these standalone blocks, an interposer is required (Figure 4).

Throughout the design process, the specific requirements and specifications for each block need to be shared and discussed. All the necessary information is gathered diligently and meticulously analyzed the options to determine the most suitable PCB substrate and stack-up configuration. The objective was to ensure the interposer's compatibility with the desired frequency band and impedance matching requirements.

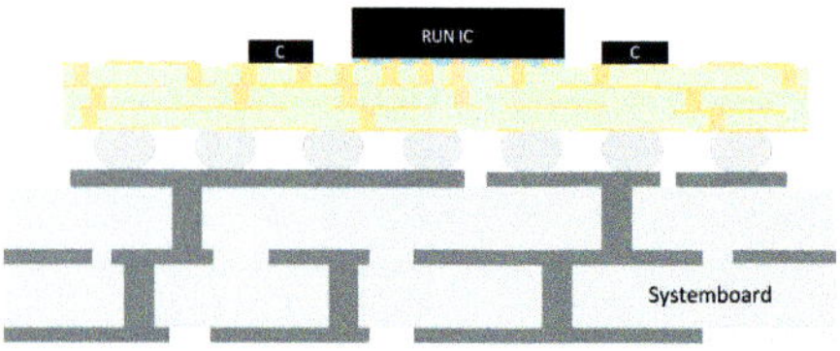

(a)

(b)

Fig. 4. Packaging concept for standalone block RFIC. (a) Cross section, (b) top view.

This work focuses on the design and the implementation of the interposer packaging, which ensures the seamless integration and successful characterization of the individual blocks within the standalone block RFIC. This collaborative process fosters a deeper understanding of the RFIC performance and facilitates the refinement of subsequent designs and development efforts.

Interposer Design Challenges

A comprehensive exploration was undertaken to investigate various configurations of interconnects, both vertical and planar, aligning with the proposed packaging concept. This investigation encompassed different types of interconnects, including stripline, microstrip, and coplanar configurations.

The focus was on understanding the advantages, limitations, and performance characteristics of each interconnect type within the proposed packaging concept. Factors such as signal integrity, impedance control, power delivery, and electromagnetic interference were considered during the evaluation process.

In addition to the interconnect configurations, the investigation also delved into the different types of vias, including blind, buried, and through-hole vias. Each via type offers unique advantages and challenges concerning signal propagation, routing, and manufacturing complexity. Evaluating these options allowed for the selection of the most suitable via type based on the specific requirements of the packaging concept.

Furthermore, the investigation encompassed the analysis of BGAs, a packaging technology widely used in modern electronic systems. The characteristics and benefits of BGAs were carefully examined, including their ability to provide high-density interconnections, robust mechanical stability, and efficient thermal dissipation [7][8][9].

By conducting this comprehensive investigation, a detailed understanding of the interconnect options, vias, and BGAs was achieved within the context of the proposed packaging concept. The insights gained from this analysis will inform the decision-making process, leading to the selection of the most appropriate interconnect, via, and BGA configurations to ensure optimal performance, reliability, and manufacturability of the packaging solution.

A meticulous investigation was carried out to assess the impact of process tolerances during the manufacturing of PCB structures. The focus of this investigation was on comparing the electrical performance of PCBs with varying levels of process tolerances.

During PCB manufacturing, certain process variations and tolerances can occur, such as variations in trace width, spacing, layer alignment, and material properties (Figure 5). These variations can have a significant influence on the electrical performance of the PCB, including signal integrity, impedance control, and overall circuit functionality.

To understand the effects of these process tolerances, a systematic comparison was conducted. PCBs with different levels of process tolerances were manufactured, keeping all other design parameters constant. The electrical performance of these PCBs was then thoroughly evaluated and compared, considering factors such as signal quality, signal loss, crosstalk, and overall circuit performance [10].

By analyzing the electrical performance of PCBs with varying process tolerances, valuable insights were gained regarding the sensitivity of the circuit to these variations. This investigation helped identify

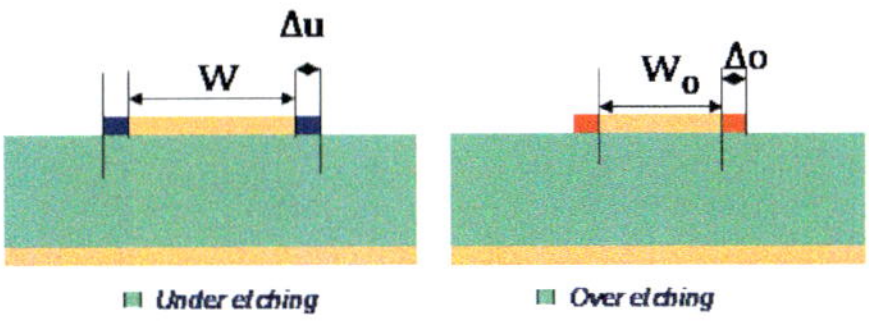

Fig. 5. Impact of etching tolerances on signal performance.

critical process tolerances that could have a substantial impact on the desired electrical performance (Figure 6).

Based on the findings, appropriate strategies can be implemented to mitigate the effects of process tolerances on the layout design during PCB manufacturing. This include refining design guidelines by full-wave 3D electromagnetic simulations to ensure tighter tolerances and more consistent electrical performance.

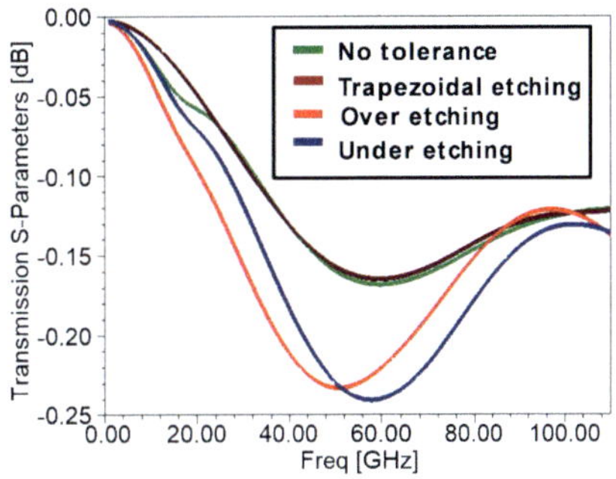

(a)

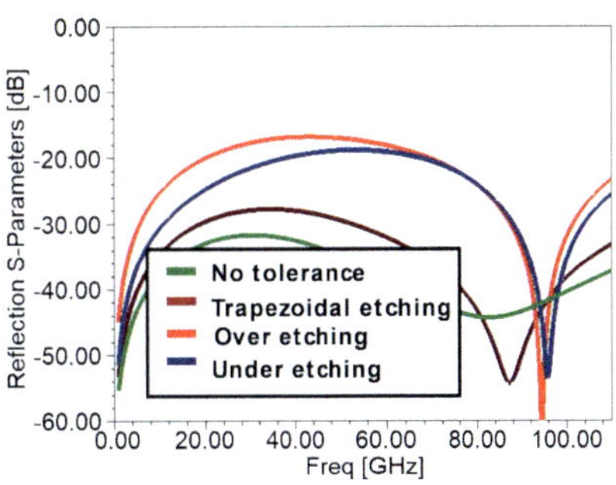

(b)

Fig. 6. Impact of etching tolerances on signal performance. (a) Transmission parameters, (b) reflection parameters.

The investigation of process tolerances during PCB manufacturing is vital for achieving reliable and robust electrical performance. By understanding the relationship between process variations and electrical characteristics, manufacturers can refine their processes and enhance the overall quality and performance of PCB structures.

Assembly of RFIC on an Organic Interposer

In order to establish the balling and assembly process of the RFIC on organic interposer boards, a method involving the use of "dummy" boards with identical top and bottom layer designs and metallization for the assembly pads was implemented. This approach allowed for testing the flux cleaning process after soldering, not only for the RFIC assembly but also for the functional boards. The cleaning step was crucial to eliminate any residues of flux that might be present underneath the dipped flip-chip, with particular attention given to the jetted solder paste utilized for surface mount device (SMD) components. The main challenge in this context was to effectively remove various flux residues both below the soldered flip-chip and from the SMD soldering process.

A comprehensive process involving balling, flip chip and SMD assembly, soldering, and cleaning was developed, assessed, and validated using the dummy PCBs as depicted in Figure 7.

The established procedure encompassed the entire manufacturing cycle, from the initial balling of the components to the precise placement and attachment of the flip chips and SMDs on the dummy PCBs. Through rigorous evaluation and verification, the effectiveness and reliability of each step in the process were thoroughly examined.

The balling process ensured the proper formation of solder balls on the components, enabling successful bonding to the PCB surface. This step was crucial for achieving strong and reliable electrical connections.

The flip chip and SMD assembly stages involved the precise positioning of the components onto the dummy PCBs, ensuring accurate alignment with the corresponding solder pads. Careful attention was given to secure and optimize the solder joints to guarantee excellent mechanical and electrical performance.

Soldering, a critical phase in the process, involved the controlled application of heat to melt the solder, allowing it to bond the components to the PCB. The temperature, duration, and soldering technique were meticulously evaluated and adjusted to achieve consistent and high-quality solder joints.

Once the assembly and soldering stages were completed, the cleaning process was initiated to remove any flux residues or contaminants that may have accumulated during the previous steps. Thorough cleaning was essential to ensure the reliability and longevity of the assembled PCBs, preventing any potential performance degradation or electrical issues [11].

Throughout the evaluation and verification of this comprehensive process, the dummy PCBs served as a reliable testing platform, allowing for iterative improvements and adjustments. The goal was to establish a robust and efficient manufacturing procedure that could be seamlessly applied to functional PCBs, ensuring consistent and reliable performance in real-world applications.

The successful assembly and cleaning of the top side, which included the SMDs and flip-chip components, validated the developed process, indicating

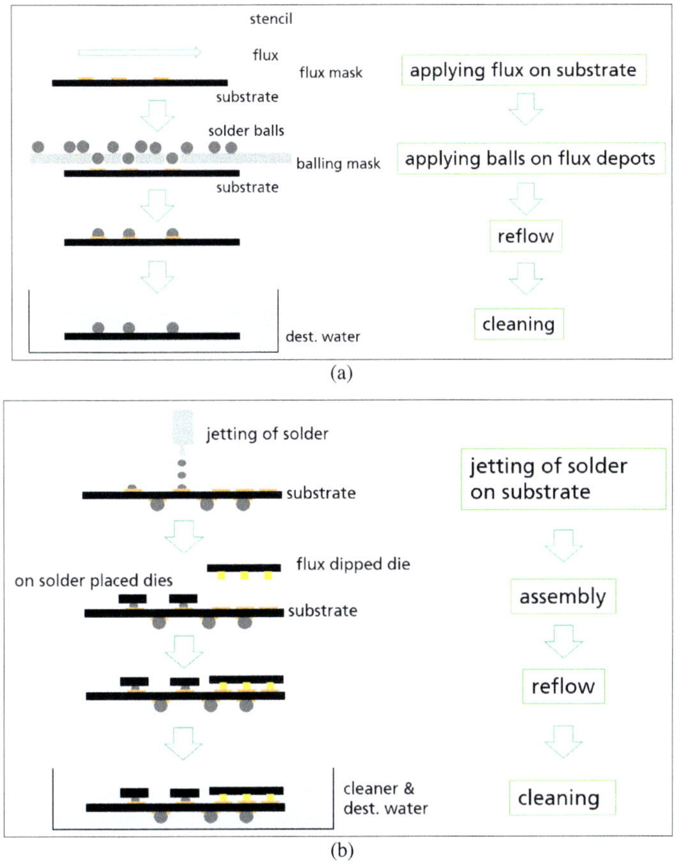

Fig. 7. (a) Schematic of balling process of the interposer bottom side, (b) schematic of assembly process of the interposer top side.

that it could be implemented without any modifications.

The fully assembled interposers underwent a sorting process and were subsequently delivered to the project partners for comprehensive functional testing. This step ensured that the interposers met the required performance and functionality criteria.

In addition to the assembly and cleaning steps, the process was enhanced by incorporating underfilling for the flip-chip components. Initially, it was planned to introduce underfilling after the first electrical test due to potential impacts on the RF performance. However, considering concerns related to reliability and stability, the decision was made to integrate underfilling into the process right from the beginning (as illustrated in Figure 8) [12][13].

By integrating underfilling early on, the process aimed to enhance the overall reliability and stability of the flip-chip components. This step addressed potential risks and mitigated any adverse effects on the RF performance, ensuring that the interposers met the required standards for long-term reliability and performance.

The decision to include underfilling from the beginning showcased a proactive

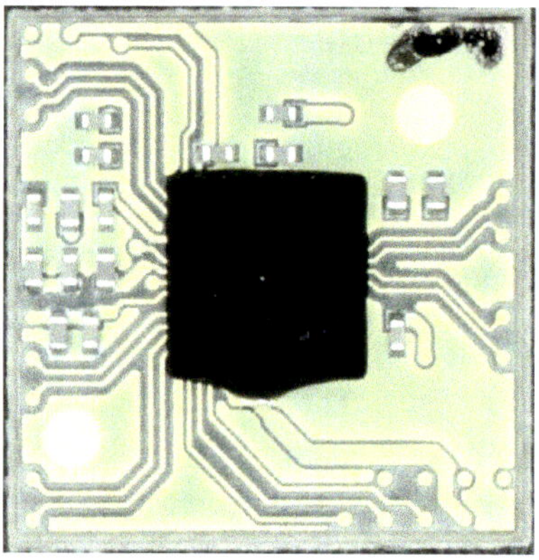

Fig. 8. Microscopic image of an underfilled MMIC.

approach towards optimizing the manufacturing process, prioritizing reliability, and streamlining the overall production workflow.

IV. DISCUSSION

The process of designing and manufacturing the interposer, as well as the assembly of RFICs, described in this paper, involves several crucial steps (Figure 9). These steps include identifying the critical signal paths, conducting RF electromagnetic simulations to minimize losses within the interposer (Figure 9a), determining the layout of the interposer, fabricating the interposer (Figure 9b), performing reflow soldering (Figures 9c–d), and conducting quality control using X-ray imaging (Figure 9e).

To begin with, it is essential to identify the critical signal paths within the interposer. These paths carry important RF signals and require special attention to ensure optimal performance. By identifying these paths, designers can focus on optimizing their layout and ensuring their integrity throughout the manufacturing process.

RF electromagnetic simulations play a significant role in minimizing losses within the interposer. Through these simulations, the behavior of RF signals can be analyzed, identify potential areas of signal degradation or interference, and make necessary adjustments to reduce losses. The goal is to optimize the interconnects and minimize any signal degradation or attenuation, ensuring efficient signal transmission.

Once the RF simulations are completed, the layout of the interposer is determined. This includes arranging the critical signal paths, incorporating necessary components and structures, and ensuring proper electrical and mechanical connectivity. The layout is designed to meet specific electrical and thermal performance requirements, considering the overall system architecture and integration constraints.

The next step involves manufacturing the interposer based on the finalized layout. The fabrication process includes the precise deposition of materials, such as the interconnects and dielectric layers, as well as the formation of vias and other necessary structures. The manufacturing process follows strict quality control measures to ensure accuracy, reliability, and compatibility with the intended application.

After the interposer is fabricated, reflow soldering is performed to connect the RFIC to the interposer. This process involves carefully aligning the RFIC and the interposer, applying solder paste, and subjecting the assembly to controlled heating, causing the solder to melt and form reliable electrical connections.

To ensure the quality and integrity of the interposer assembly, X-ray imaging is commonly employed for quality control. X-ray images allow for non-destructive inspection, allow to verify the solder joints, detect any potential defects or anomalies, and ensure the overall quality of the assembly.

The simulation, fabrication, and assembly of the interposer, coupled with quality

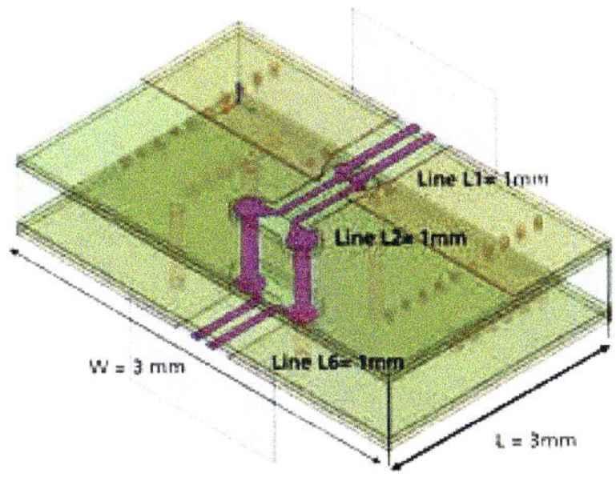

(a)

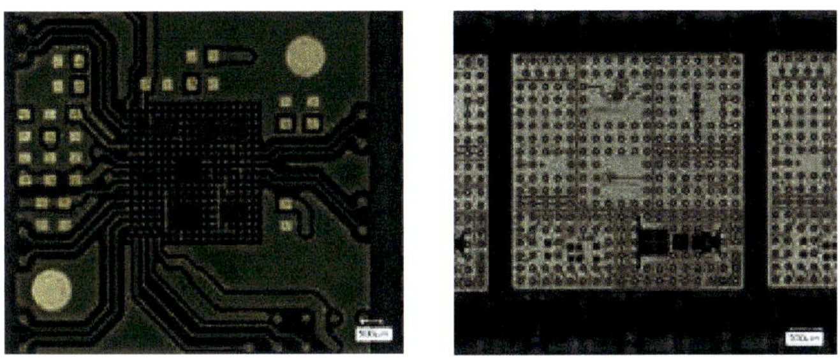

(b) Left: top view, right: bottom view

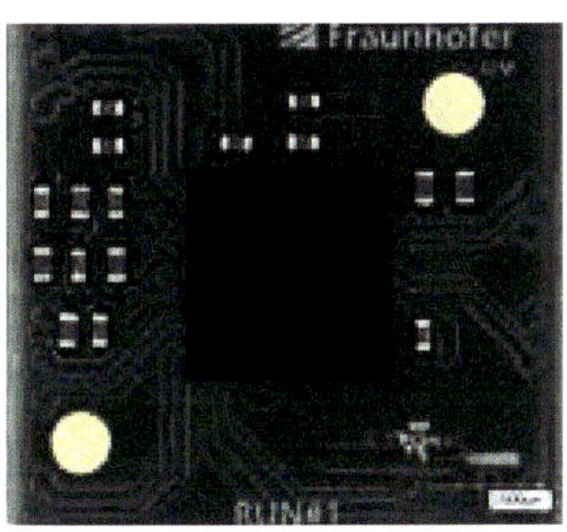

(c) Top view

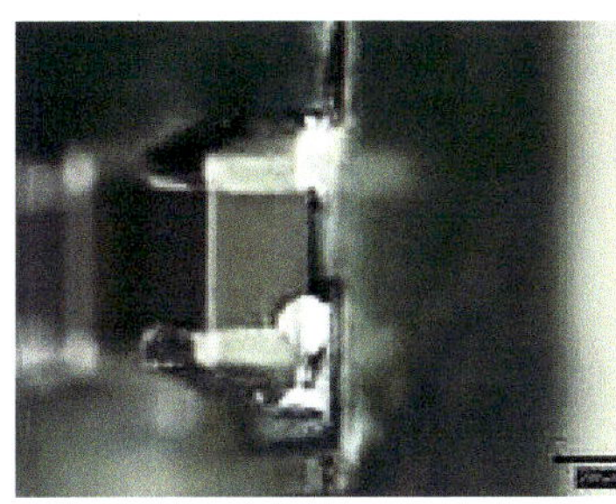

(d) Side view

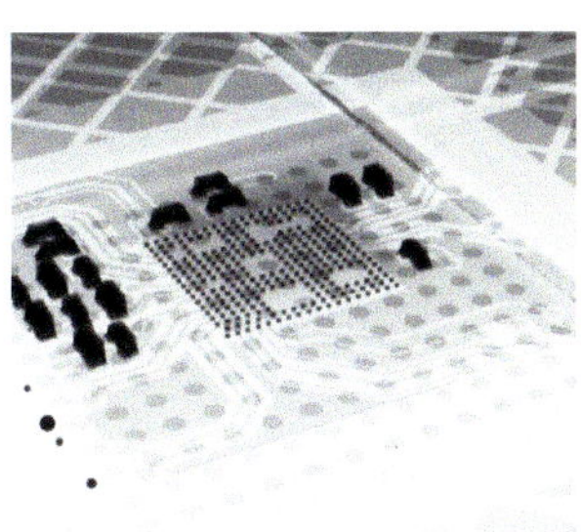

(e)

Fig. 9. (a) RF paths simulations, (b) manufactured interposer, (c) after reflow soldering, (d) soldered capacitor, (e) X-ray image for solder quality control.

control measures, ensure the production of a high-performance and reliable interposer platform for integrating RFICs. By carefully addressing critical signal paths, conducting RF simulations, optimizing the layout, and employing quality control techniques, the design and manufacturing process ensures the interposer's functionality, performance, and adherence to the required standards.

V. CONCLUSION AND PERSPECTIVES

The packaging technology employed in front-end components is a pivotal aspect of designing 5G communication systems, primarily due to its vital role in supporting high-frequency and high-power requirements. This technology is instrumental in achieving the miniaturization, integration, and reliability demanded by these components. This article aims to present a comprehensive workflow for designing the packaging technology specific to 5G communication systems, as part of the BEYOND5 project. It has covered key components, integration challenges, and the use of electromagnetic simulations.

The design process for an interposer, a critical component, begins with the careful selection of material. The choice of material is primarily driven by the need to tolerate losses at high frequencies effectively. Once the suitable material is identified, other factors such as the bump pitch on the front-end chip and the complexity of routing on the interposer come into play, influencing the selection of the appropriate stack-up configuration.

The purpose of developing this standalone block RFIC interposer is to facilitate the measurement of the high-speed blocks within the 5G demonstrator transceiver. By enabling isolated testing of these specific blocks, valuable information can be obtained to guide the development of the final demonstrator.

The next phase involves the development of an interposer specifically designed for a single transceiver RFIC. This interposer will serve as a crucial intermediate step, providing valuable insights and data that will inform the design and construction of the final 5G demonstrator. The knowledge and experience gained from testing the single RFIC interposer will contribute to the refinement and optimization of the final interposer for the in-flight communication demonstrator.

By leveraging the same technology and approach used in developing the standalone block RFIC interposer, the demonstrator interposer can be efficiently created. This ensures a seamless transition and maximizes the utilization of resources and expertise. The successful implementation of the standalone block RFIC interposer will pave the way for a robust and effective interposer solution in the final 5G demonstrator, enabling comprehensive testing and validation for in-flight communication capabilities.

REFERENCES

[1] A. Gupta and R. K. Jha, "A survey of 5G network: Architecture and emerging technologies," in *IEEE Access*, vol. 3, pp. 1206-1232, 2015, doi: 10.1109/ACCESS.2015.2461602.

[2] H. Ito et al., "Advanced low-loss and high-density photosensitive dielectric material for RF/millimeter-wave applications," *2019 International Wafer Level Packaging Conference (IWLPC)*, San Jose, CA, USA, 2019, pp. 1-6, doi: 10.23919/IWLPC.2019.8914136.

[3] Y. P. Zhang and D. Liu, "Antenna-on-chip and antenna-in-package solutions to highly integrated millimeter-wave devices for wireless communications," in *IEEE Transactions on Antennas*

and Propagation, vol. 57, no. 10, pp. 2830-2841, Oct. 2009, doi: 10.1109/TAP.2009.2029295.

[4] T. Thai, S. Dalmia, J. Hagn, P. Talebbeydokhti and Y. Tsfati, "Novel multicore PCB and substrate solutions for ultra broadband dual polarized antennas for 5G millimeter wave covering 28GHz & 39GHz range," *2019 IEEE 69th Electronic Components and Technology Conference (ECTC)*, Las Vegas, NV, USA, 2019, pp. 954-959, doi: 10.1109/ECTC.2019.00149.

[5] A. O. Watanabe, M. Ali, S. Y. B. Sayeed, R. R. Tummala and M. R. Pulugurtha, "A review of 5G front-end systems package integration," in *IEEE Transactions on Components, Packaging and Manufacturing Technology*, vol. 11, no. 1, pp. 118-133, Jan. 2021, doi: 10.1109/TCPMT.2020.3041412.

[6] I. Ndip and K.-D. Lang, "Roles and requirements of electronic packaging in 5G," *7th Electronic System-Integration Technology Conference (ESTC)*, 2018, doi: 10.1109/ESTC.2018.8546469.

[7] Y. Kawano *et al.*, "Flip chip assembly for sub-millimeter wave amplifier MMIC on polyimide substrate," *2014 IEEE MTT-S International Microwave Symposium (IMS2014)*, Tampa, FL, USA, 2014, pp. 1-4, doi: 10.1109/MWSYM.2014.6848323.

[8] W. Heinrich, "Flip-chip for millimeter-wave and broadband packaging," *2005 IEEE International Wkshp on Radio-Frequency Integration Technology: Integrated Circuits for Wideband Comm & Wireless Sensor Networks*, Singapore, 2005, pp. 124-126, doi: 10.1109/RFIT.2005.1598871.

[9] T. Braun, K.-F. Becker, M. Koch, V. Bader, R. Aschenbrenner, H. Reichl "Flip chip technology for high temperature automotive applications" *Proc. of ECTC 2005, Orlando, Fl., USA.*

[10] K. S. Murugesan et al., "Modeling and measurement of double stacked microvia in antenna-in-package module for 5G mmWave applications," *2021 51st European Microwave Conference (EuMC)*, London, United Kingdom, 2022, pp. 10-13, doi: 10.23919/EuMC50147.2022.9784283.

[11] C. Dominkovics and G. Harsanyi, "Effects of flux residues on surface insulation resistance and electrochemical migration," *2006 29th International Spring Seminar on Electronics Technology, St. Marienthal, Germany*, 2006, pp. 206-210, doi: 10.1109/ISSE.2006.365387.

[12] J. Pyland, R. Pucha and S. Sitaraman, "Effect of underfill on BGA reliability," *Proceedings International Symposium on Advanced Packaging Materials Processes, Properties and Interfaces (IEEE Cat. No.01TH8562), Braselton, GA, USA*, 2001, pp. 354-359, doi: 10.1109/ISAOM.2001.916601.

[13] K.-F. Becker, N. Kilic, T. Braun, M. Koch, V. Bader, R. Aschenbrenner, H. Reichl, "New insights in underfill flow and flip chip reliability," *Proc. Apex 2003; 29.03.-02.04.03, Anaheim, Ca., USA.*

Tekfouy Lim received his B.Sc. degree in electrical and electronics engineering and M.Sc. degree in radio frequency and millimeter-wave engineering from Lille 1 University, in France, in 2007 and 2009, respectively, and his Ph.D. degree from the University of Grenoble in France (in collaboration with STMicroelectronics, Crolles, France and the IMEP-LAHC Laboratory, Grenoble, France), in 2013. Between 2013 and 2018, he was a Research Associate with the Fraunhofer Research Institution for Microsystems and Solid State Technologies EMFT, in Munich, Germany. Since 2020, he has been part of the Fraunhofer Institute for Reliability and Microintegration IZM, in Berlin, Germany. His main research interests involve the packaging integration of millimeter-wave integrated circuits for telecommunication applications.

Michael Kaiser received his B.Sc. and M.Sc. in electrical engineering and information technology from Technical University Munich, Germany in 2014 and 2017, respectively. Since 2017, he has been working as a research scientist and group manager in the department RF & Smart Sensor Systems at Fraunhofer – Institute for Reliability and Micro-integration (IZM), Germany. His research interests include hardware architectures and integration platforms for energy-efficient mm-wave and (sub-) THz front-end modules for wireless communication applications.

Mattis Obst studied microsystem technologies at the University of Applied Sciences in Berlin and joined Fraunhofer IZM in 2016. In 2017 he received his Bachelor's degree and in 2020 his Master's degree, both focused on dielectric breakdown of epoxy resins and the applied measuring routines. Mattis Obst is member of the group Assembly & Encapsulation Technologies. Recent research has been focused on fan-out wafer and panel level packaging technologies and Mattis Obst takes part in the Fan-out Panel Level Packaging Consortium at Fraunhofer IZM Berlin.

Tanja Braun studied mechanical engineering at Technical University of Berlin with a focus on polymers and micro systems and joined Fraunhofer IZM in 1999. In 2013 she received her Dr. degree from the Technical University of Berlin for the work focusing on humidity diffusion through particle-filled epoxy resins. Tanja Braun is head of the group Assembly & Encapsulation Technologies. Recent research is focused on fan-out wafer and panel level packaging technologies. She is an active member of IEEE. She is a member of the IEEE EPS Board of Governor (BOG) and the IEEE EPS Region 8 Program Director.

Marius van Dijk received his M.Sc. degree in mechanical engineering from the Eindhoven University of Technology in 2011. Since 2013 he has been working as a scientific researcher at the Fraunhofer Institute for Reliability and Microintegration IZM in Berlin, Germany. The main focus of his work is performing numerical simulation (finite element methods) of thermal, thermomechanical, thermal-electrical, diffusion and multi-physics models. Additionally, a special focus of his research is on modelling ageing effects of polymeric materials used in microelectronic applications.

Johannes Jaeschke completed his doctoral thesis on "failure mechanism electromigration in solder joints" in 2012 after studying electrical engineering at Technische Universität Berlin. Since 2015 he has been Group Manager of the group 'System Reliability Assessment' of IZM's department Environmental and Reliability Engineering. His research focus is micro system reliability, application specific condition monitoring, reliability testing. Furthermore he supervises the scientific work of BSc, MSc and PhD students at IZM as well as part of the Research Center for Microperipheric Technologies at Technische Universität Berlin. He is a University lecturer at Technische Universität teaching courses with the focus on microsystem technologies and reliability.

Lars Böttcher received in 2000 an engineering degree in micro-system technologies from the University of Applied Science and Business (Berlin, Germany) and is currently an R&D engineer and project manager at Fraunhofer Institute for Reliability and Micro Integration (IZM) in Berlin, were he leads the group "Embedding and Substrate Technologies".

He has worked in the Department for System Integration and Interconnection Technologies for 23 years, and he specializes in packaging process development with emphasis on printed circuit board manufacturing technologies. He is responsible for different projects with industrial partners, as well as German and European Union funded projects, which focus on new package technologies based on embedded chips and new substrate technologies.
He is a member of the technical committee for the SMTA International conference and technical chair for the IMAPS Device Packaging conference.

Kavin Senthil Murugesan received his Bachelor degree in electrical and electronics engineering (2014) from P.A. College of Engineering and Technology, Anna University, India. He completed his Masters in Nanoelectronic Systems in 2017 from Technical University Dresden, Germany. Since 2018, he has been working as a research scientist in the department of Research focus technologies of Microperipherics, Technical University Berlin in close cooperation with Fraunhofer Institute for Reliability and Microintegration (IZM), in Berlin, Germany. His main research interests involve packaging interconnects, and passive components for mm-wave and (sub-) THz front-end modules for wireless applications.

Uwe Maass holds a Master's degree in Electrical Engineering from Technische Universität Berlin, Germany. He is currently with Fraunhofer IZM as research assistant. He has been involved with RF package design and characterization and electromagnetic modelling of components for RF applications. His current research interests includes antenna-in-package design for mmWave frequencies.

Ivan Ndip (M'05–SM'12) received his Dipl.-Ing. (M.Sc.) and Dr.-Ing. (Ph.D.) degrees (summa cum laude) in electrical engineering from the Technical University (TU) of Berlin, Germany, in 2002 and 2006, respectively, and his Dr.-Ing. habil. Degree, also in electrical engineering, from the Brandenburg University of Technology, Cottbus-Senftenberg, Germany, in 2017. He joined the Fraunhofer Institute for Reliability and Microintegration, IZM, Berlin, as a Student Research Assistant in 2000 and a Research Engineer in 2002. Since 2014, he has been the Head of the Department of RF and Smart Sensor Systems, IZM, where he leads research and development activities in five research groups and manages the department. He has been a Lecturer with the School of Electrical Engineering and Computer Sciences, TU of Berlin since 2008. He also teaches Professional Development Courses to practicing engineers and scientists worldwide. He has authored or co-authored over 175 publications in referred journals and conference proceedings. He is an inventor and holds many patents and patent applications. Dr. Ndip is a member of the Technical Program Committee of many IEEE and IMAPS international conferences. He is also a reviewer for IEEE Transactions on Electromagnetic Compatibility, IEEE Transactions on Components, Packaging, and Manufacturing Technology, IEEE Transactions on Microwave Theory and Techniques, IEEE Transactions on Electron Devices, and other international journals. He is an Associate Editor of the Journal of Microelectronics and Electronic Packaging.

Energy Efficient Beam Control for 5G Antennas

Damian Duraj, Luiza Leszkowska, Weronika Kalista, Kamil Trzebiatowski, Lukasz Kulas, and Krzysztof Nyka

Abstract—**The rapid development of 5G and beyond systems demands improvement in communication speed, latency and safety to maintain the required quality of service. This paper presents an overview of different concepts of energy-efficient antenna systems, which offer beam-shaping and beam-steering functionalities, that enhance connectivity performance and can be used in 5G applications. Different designs for 5.9 GHz, 39 GHz and 60 GHz frequency bands are proposed to further improve the capabilities of 5G MIMO systems or reduce energy consumption, i.e. antennas with low-power CMOS switches, antenna arrays that can be integrated with 3D-printed lenses, patch antennas with lenses, antennas with planar reconfigurable superstrate to focus the beam and a multi-beam antenna based on a multiport patch array producing beams deflected by an integrated 3D-printed lens.**

This paper is a result of the BEYOND5 (www.beyond5.eu) project which has received funding from the ECSEL Joint Undertaking (JU) under grant agreement No 876124. The JU receives support from the European Union's Horizon 2020 research and innovation programme and France, Germany, Turkey, Sweden, Belgium, Poland, Netherland, Israel, Switzerland, and Romania.

D. Duraj, L. Leszkowska, W. Kalista, K. Trzebiatowski, L. Kulas, and K. Nyka are with the Department of Microwave and Antenna Engineering, Faculty of Electronics, Telecommunications and Informatics, Gdansk University of Technology, Gdańsk, Poland (e-mail: {damian.duraj, luiza.leszkowska, weronika.kalista, kamil.trzebiatowski, lukasz.kulas, krzysztof.nyka} @pg.edu.pl)

Index Terms—**5G, switched-beam antenna, reconfigurable antenna, beam control, lens antenna, antenna array, direction-of-arrival estimation, internet-of-things, wireless sensor networks, ESPAR antenna, ERES antenna, 3D printing.**

I. INTRODUCTION

With the rapid proliferation of wireless communication and the advent of fifth-generation (5G) technology, the demand for faster data rates, higher capacity, and seamless connectivity has rapidly risen. 5G networks are envisioned to revolutionize various industries, enabling advancements in autonomous vehicles, smart cities, and the Internet of Things (IoT). However, the massive scale and bandwidth requirements of 5G networks pose significant challenges, especially in terms of energy consumption. Energy efficiency has emerged as a critical concern in the design and deployment of 5G networks as well as future systems. Traditional cellular networks already account for a non-negligible portion of global energy consumption, and the exponential growth of 5G is expected to further amplify this issue.

Among the key elements of a 5G infrastructure, beamforming technology plays a pivotal role in optimizing the performance of antenna systems. Beam control techniques enable the concentration of radio frequency (RF) energy towards specific users or areas of interest, thereby enhancing signal quality, reducing interference, and improving network capacity. However, different beam control methods have different trade-offs between performance and complexity, and require different antenna designs and hardware implementations.

One of the key technologies considered in the field of antennas is phased arrays combined with digital beamforming that allow for flexible spatial diversity in challenging multi-user scenarios [1]. Being the most versatile technique, it enables the use of popular and well-established technologies that improve the communication efficiency, such as orthogonal frequency-division multiplexing (OFDM), multiple-input-multiple-output (MIMO), and massive MIMO [2]-[4]. However, as it requires multiple transceiver (TxRx) units and powerful digital signal processing (DSP), digital beamforming cannot be used easily and efficiently in all possible 5G applications due to high complexity, costs, and energy consumption.

To enable new applications, in which costs or energy consumption can play an important role, reconfigurable antennas have been proposed [5],[6]. Different approaches to the beam control are possible, including reconfigurable fixed beam antennas, switched beam antennas and multibeam antennas. In the first group of antennas, the beam shape and direction is changed by adding replaceable dielectric lenses (lens antennas) [7] [8] or planar structures comprising arrays of parasitic conductive patches (superstrate antennas) [9]. By carefully designing the lens geometry and materials, lens antennas can achieve high directivity and gain, enabling efficient beamforming with reduced power consumption. Similar effects can be achieved in superstrate antennas. In both these techniques, the radiating beam is usually narrowed to increase antenna gain and provide improved spatial diversity.

Switched-beam antennas have garnered significant attention as a mean of improving energy efficiency in 5G systems. By dynamically switching between multiple fixed radiation patterns, these antennas can adapt to the spatial distribution of users and direct the RF energy more precisely, thus reducing wasted energy. They can improve the signal-to-noise ratio (SNR) and reduce the interference by selecting the best beam for each user or direction [10],[11]. However, switched-beam antennas have limited flexibility and adaptability, as they cannot steer or shape the beams in 2D continuously. There are two main techniques used for beam switching: multiple switched radiating elements generating different beams or a single radiating element in which radiation is modified by an array of switched parasitic elements, e.g. electronically steerable parasitic array radiator (ESPAR) antennas [12]. The first group of switched beam antennas can be immediately adapted to multi-beam antennas that represent another promising approach for energy-efficient beam control in 5G networks [13]. These antennas employ multiple radiating elements to generate multiple simultaneous beams of predefined shapes and directions, allowing for efficient allocation of RF energy to multiple users or sectors.

This paper presents an overview of different concepts and designs of antenna systems developed during the BEYOND5 project that offer energy-efficient beam-shaping and beam steering functionalities in a wide set of frequency ranges from 5.9 GHz up to millimeter waves at 39 GHz and 60 GHz. We have focused on the beam-control techniques that utilize reconfigurable fixed beam antennas, switched beam antennas and multibeam antennas. The proposed designs can easily be produced in a low-cost process and applied in practical 5G systems.

II. ANTENNA DESIGNS

5.9 GHz ESPAR Antenna

A number of electronically steerable parasitic array radiator (ESPAR) antenna for the 2.4 GHz frequency band has been reported previously [12],[14]. Unlike other solutions that require a complex energy-intensive beamforming network integrated with an array, the available concepts offer beam-steering capabilities that can be controlled with a digital input-output (DIO) port from a basic microcontroller. Such simple control mechanism ensures a seamless integration with complex algorithms.

The ESPAR antenna concept was successfully adapted to vehicle-to-everything (V2X) communication in the 5.9 GHz frequency band. The complete description of the suggested antenna design can be found in [15]. A model of the antenna, showing both its top and side views, is presented in Figure 1. The proposed antenna consists of a single SMA connector acting as an active monopole surrounded by twelve parasitic monopoles connected to switching circuits which alters their impedance between two states. The base for the antenna ground plane and switching circuits is an inexpensive FR4 substrate. The switching circuits have been realised using the energy-efficient PE42424 single-pole double-throw (SPDT) switch. The simplicity of the design enables the definih tion of a steering configuration as a vector representing the state of each parasitic element with "0" or "1" logical states and offers a broad range of radiation patterns from omnidirectional to directional with a gain equal to 5.8 dBi. The designed antenna model underwent validation using the power pattern cross-correlation (PPCC) algorithm. As shown in Figure 2, the ESPAR antenna and the PPCC algorithm for direction-of-arrival estimation (DoA) provide the estimation error below 7° for three steering vectors, which produce unique main beams that can be rotated in the horizontal plane, that were obtained [16].

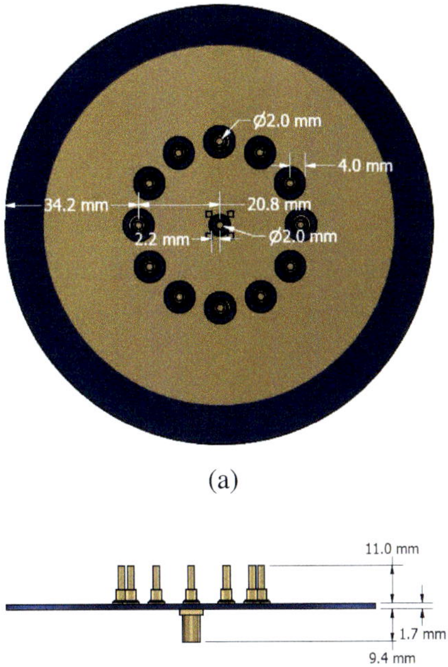

(b)

Fig. 1. V2X ESPAR antenna model: (a) top view, (b) side view.

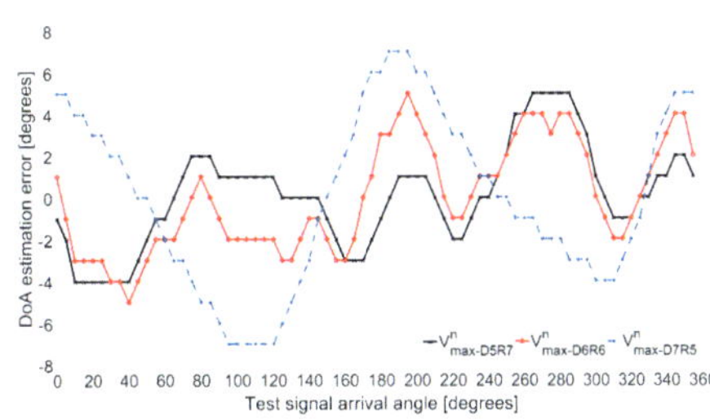

Fig. 2. V2X ESPAR antenna DoA error estimation for three directional radiation patterns that can be rotated in the horizontal plane [16].

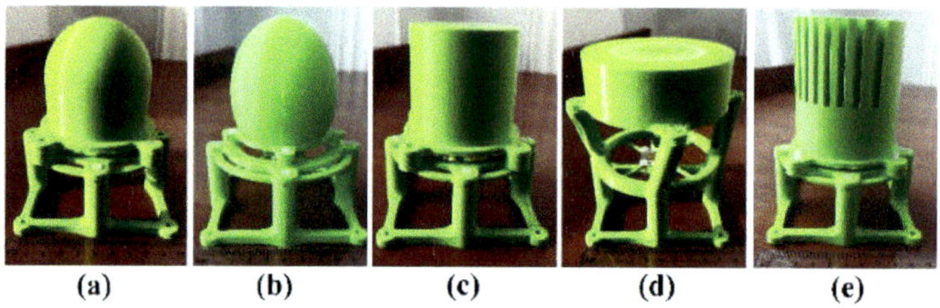

Fig. 3. Prototypes of antenna with lenses (antenna dimensions $D \times H$): (a) extended hemispherical (70 × 86.1 mm), (b) ellipsoidal lens (70 × 91.0 mm), (c) cylindrical lens (70 × 88.1 mm), (d) GRIN lens (100 × 88.1 mm), (e) polarizer lens (73.4 × 101.1 mm).

5.9 GHz Patch Antenna with Integrated Lenses

The proposed patch antenna with integrated lenses is a set of single beam lens antennas for 5.9 GHz with the same radiator and exchangeable 3D printed lenses. Using different lenses it is possible to shape the beam as is required by a specific application requirements. A simple rectangular coaxially fed patch on RO3003C laminate was used as a radiator and five different lenses were designed and produced. The basic function of all the lenses was beam narrowing resulting from gain enhancement, while one lens performed the function of a wave polarizer as well. For the sake of comparison, the lenses were fabricated using the same low-cost PLA material (measured $\varepsilon_r = 2.63$ and $\tan\delta = 0.01$) and, if possible, a similar base diameter (about 70 mm) was preserved. Pictures of manufactured antenna prototypes are presented in Figure 3. The design considerations steps were reported in [17] and [18], while not reported was that their cylindrical lens was optimized as a cut conical shape. The realised gain of antennas achieved a minimum of 12.9 dBi and a maximum of 14.9 dBi in the case of extended hemispherical and cylindrical lenses, respectively, compared with 7.7 dBi gain of the patch. This significantly exceeds the gain of directional antennas found in V2X systems, such as 8.76 dBi [19] and around 10 dBi in [20], both systems utilizing patch antennas. However, our antennas were designed to verify the extent to which we could further improve connectivity by increasing the gain, as even replacing omnidirectional antennas with moderate gain antennas provided significant improvements. All of the measured results of prototypes characterized in an anechoic chamber are compared in Figure 4 and Table 1. The results show that by a simple change in antenna enclosure, a

Fig. 4. Measured realised gain radiation patterns of antenna with different lenses: (a) E-plane measurement, (b) H-plane measurement.

TABLE 1
COMPARISON OF MEASURED PARAMETERS OF REALISED ANTENNA WITH LENSES

	Realised gain [dBi]	Sidelobe level (SLL) [dB] E/H-plane	3 dB beamwidth [°] E/H-plane	Polarization
Patch	7.7	–/–	72/85	LP
Extended hemispherical	12.9	−15.8/−13.6	33.6/29.9	LP
Ellipsoidal	13.9	−14.8/−12.2	33.1/30.2	LP
Cylindrical	14.9	−18.1/−13.2	29.9/27.7	LP
GRIN	13.6	−14.0/−12.2	29.5/29.3	LP
Polarizer	14.0	−12.8/−12.5	29.4/29.4	CP

single antenna design could be adapted to diverse scenarios with different requirements.

5.9 GHz Array Antenna with Lens for Multibeam Application

The proposed array antenna with lens for multibeam application is a simple, yet effective, antenna developed for the 5.9 GHz V2X frequency band. The design is based on a planar array of 2 × 2 antennas assembled with a half-ellipsoidal shape lens. Multiple independent beams can be switched by choosing the right excitation port. Due to the offset of the radiators relative to the lens base centre, beams are deflected along the lens' diagonals, in four different directions.

The array consists of four circular patch antennas fed through C-shape slots, this design is based on [21], with linear polarization, on a CuClad217 substrate. Neighbouring radiators are at a distance of 38 mm from each other. The lens comes in a shape of a half ellipsoid with a radius equal to 57 mm in the x and y axes and 51 mm in the z axis. It is kept at a distance of 16 mm from the radiator and is manufactured by 3D printing with ABS300 Preperm filament with measured dielectric properties $\varepsilon_r = 2.93$ and $\tan\delta = 0.01$. The model of the antenna array and integration with the lens is shown in Figure 5.

Beams are deflected in four different directions, namely V1 to Φ: 225°; V2 to Φ: 135°; V3 to Φ: 45°; and V4 to Φ: 315°, at angle θ equal to 25° ± 2°. The 3 dB beams of the antenna can be clearly distinguished and its angular coverage is presented in Figure 6. Apart from deflecting the beam, the lens also increases the gain of the radiators and the antenna with single excitations provides moderate gain of over 12 dBi. The gain varies among beams by at most 0.5 dB and sidelobe level (SLL) remains below −13.8 dB.

5.9 GHz ERES antenna

The initial concept of the electronically reconfigurable superstrate (ERES) antenna has been described in [9], in which simulation results of such antenna were presented for the 2.4 GHz frequency band. It is a planar antenna composed

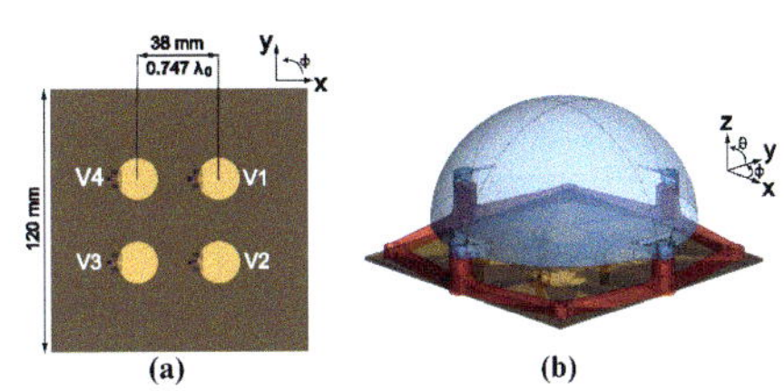

Fig. 5. (a) Antenna array model top view. (b) Antenna array with lens and spacer (antenna dimensions with lens: 120 × 120 × 82 mm).

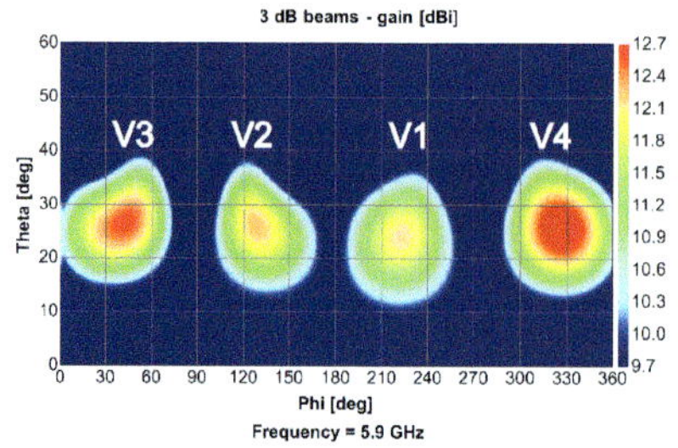

Fig. 6. Simulated 3 dB gain beams coverage at 5.9 GHz for all possible beams

of two layers (separated by air): a layer with a feed patch and a reconfigurable layer with switched passive elements. The antenna can be controlled using simple RF switches (in this case PIN diodes) connecting these passive elements to the ground plane. The states of particular RF switches can be changed using a microcontroller. Such an antenna allows for reconfiguration of the main beam in five different directions: in the direction of the *z*-axis and in four directions with a certain angle from the *z*-axis.

In this paper, we present an initial design of ERES antenna for the 5.9 GHz frequency. In this case, the simulation investigations are presented based on ideal short-circuits and ideal open-circuits on the switching layer. The ERES antenna designed for 5.9 GHz is shown in Figure 7. All the paths of the steering network are retained in the simulation model and all 16 passive elements are grouped into four switched sections, so the antenna has four switched configurations and one non-switched in the direction of the *z*-axis. The overall dimensions of the antenna are 42 × 42 × 40 mm^3 (including the coaxial connector).

In Figure 8, four possible antenna radiation patterns in the horizontal plane are presented. The direction of maximum radiation in this plane (θ = 25°) in four different configurations occurs at the following angles: $\phi1$ = 84°, $\phi2$ = 166°, $\phi3$ = 263°, $\phi4$ = 326°. In Figure 9, radiation pattern characteristics in the elevation plane are presented. For the elevation plane of ϕ = 90° presented switched configurations "0001", "0100" beams (angle of maximum radiation) deflect respectively to $\theta1$ = 27° and $\theta2$ = 28° relative to the *z*-axis, while in the elevation plane of ϕ = 0° the following configurations "0010", "1000" beams deflect respectively to $\theta3$ = 5° and $\theta4$ = 10° relative to the *z*-axis. Such asymmetry results from the asymmetry of the feeding patch relative to the switching system. Thus, further work on the development of this antenna concept will be mainly focused on balancing the symmetry of all beams. The realised gain of the antenna in the non-deflected configuration is 7.8 dBi, while the realised

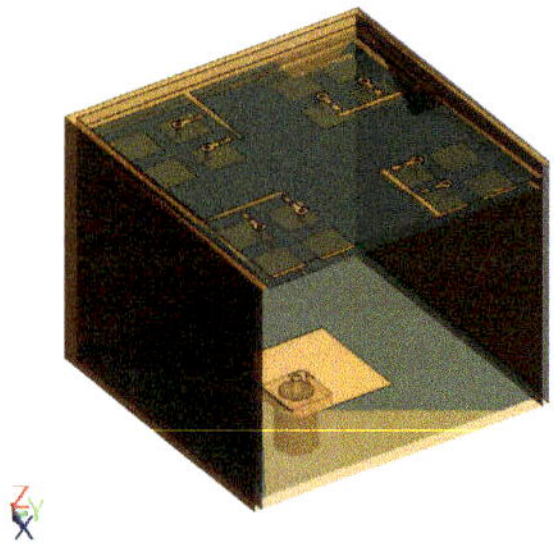

(a)

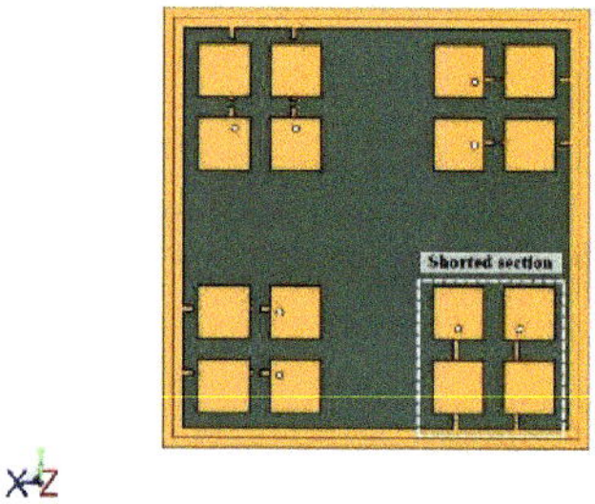

(b)

Fig. 7. ERES antenna for 5.9 GHz. (a) Antenna model with one metallic wall removed. (b) Bottom side of the reconfigurable layer (antenna dimensions: 42 × 42 × 40 mm^3).

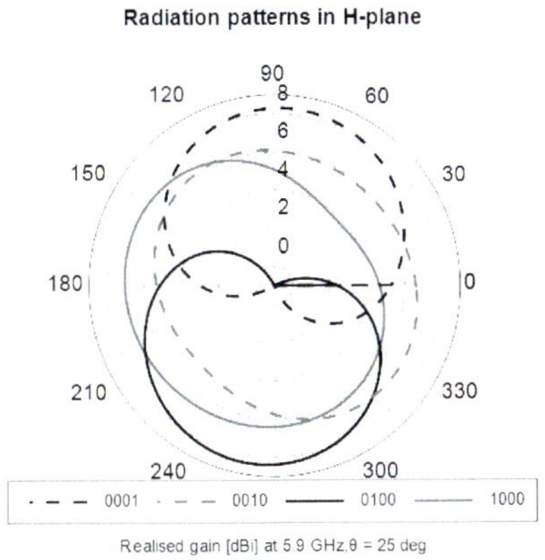

Fig. 8. Simulated radiation patterns of the ERES antenna in the H-plane.

gain of the switched configuration varies between 7.6 and 8.4 dBi.

39 GHz antenna array concept from connectorized solution to integrated multibeam with lens

An interesting concept for energy-efficient beam control working in the 39 GHz frequency band is a 4 × 4 dual-polarized microstrip antenna array, with a simplified stack-up, designed for a 5G network inside an airplane [21]. The array together with a custom radio frequency integrated circuit (RFIC) can form a highly integrated antenna array system (AAS) that can be used in a 5G base station.

The proposed antenna design is based on a circular microstrip patch antenna of dual-polarized operation. The antenna features a symmetrical shape and is fed through two orthogonal C-shaped slots on the ground plane. This two-layer design allows obtaining isolation between cross-polarized ports of a single element higher than 20 dB and cross-polarization radiation level lower than –20 dB. To facilitate the feeding process, miniature SMPS connectors are soldered to the microstrip lines located on the bottom layer. The antenna array comprises 16 elements, each consisting of a radiating patch and feeding network. The spacing between the array elements is chosen to be 0.75 times the free space wavelength (λ0) at the resonance frequency of 39 GHz. The entire antenna array is enclosed within a 3D printed package, as illustrated in Figures 10(a) and (b). A single patch can

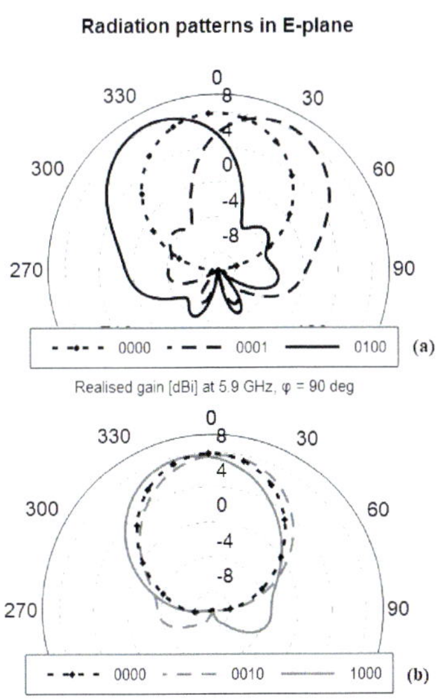

Fig. 9. Simulated radiation patterns of the ERES antenna in the E-plane: (a) ϕ = 90°, (b) ϕ = 0°.

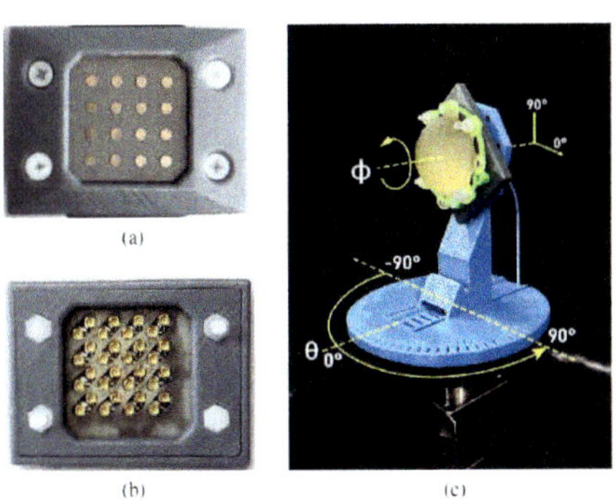

Fig. 10. Fabricated antenna: (a) array top view, (b) array bottom view, (c) array with the lens mounted on a turntable in the anechoic chamber (antenna dimensions 55 × 41 × 5 mm, with lens 55 × 41 × 26 mm (W × L × H)).

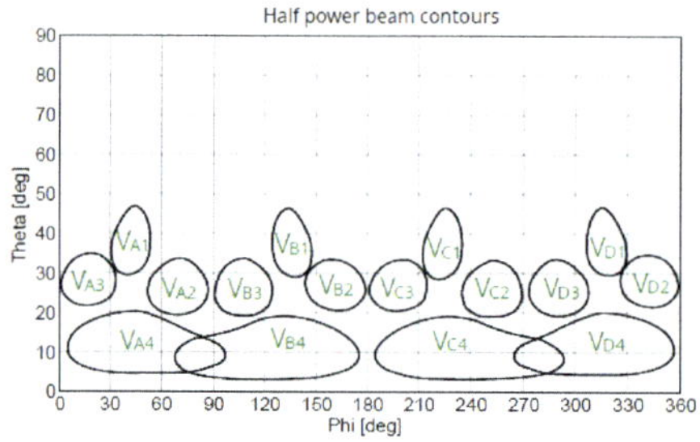

Fig. 11. Simulated half-power gain contours at 39 GHz for all possible beams.

be excited with horizontal (H) or vertical (V) polarization. The gain of a single radiator is about 7 dBi.

In order to extend the functionality of the antenna array originally designed, optimized and fabricated for stand-alone applications in 5G multi-user multiple input multiple output (MU-MIMO) systems, our next objective was to adapt it for beam-switching by integrating it with a 3D printed dielectric lens [22]. Through this modification, the array became a passive multi-beam antenna capable of radiating 16 different beams (for both polarizations) depending on the excited port. The proposed antenna can be employed for beam-switching operations with the addition of a suitable switching circuit, allowing for dynamic control of the main beam's direction with a high gain of about 15 dBi. The antenna with the lens is shown in Figure 10(c), and the 16 possible beams are presented in Figure 11. The array enclosed in the package has a size of 55 × 41 × 5 mm, and the size of the array with a dielectric lens is 55 × 41 × 26 mm.

The next step was to propose a new bottom layer design to replace an array of connectors with a ball grid array (BGA). This modification allows the integration of the antenna array with other PCBs, resulting in an antenna-in-package (AiP) solution. The design of an example of antenna-in-package stack-up is presented in Figure 12 and the detailed design of the array's BGA footprint is shown in Figure 13.

60 GHz Dual radiator reconfigurable antenna concept

Another interesting energy-efficient antenna concept developed for millimetre waves is a simple 60 GHz switched beam antenna for 5G applications [23]. The proposed design demonstrates the capability

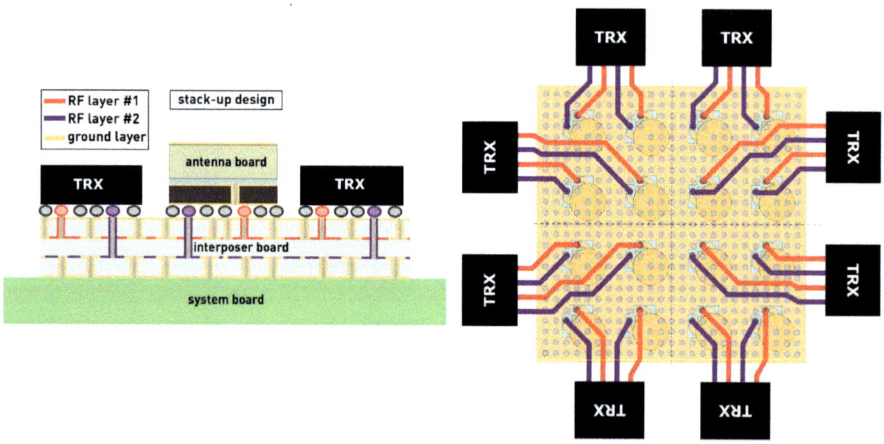

Fig. 12. The antenna-in-package stack-up and an example of connections between the microstrip patches and transceiver modules.

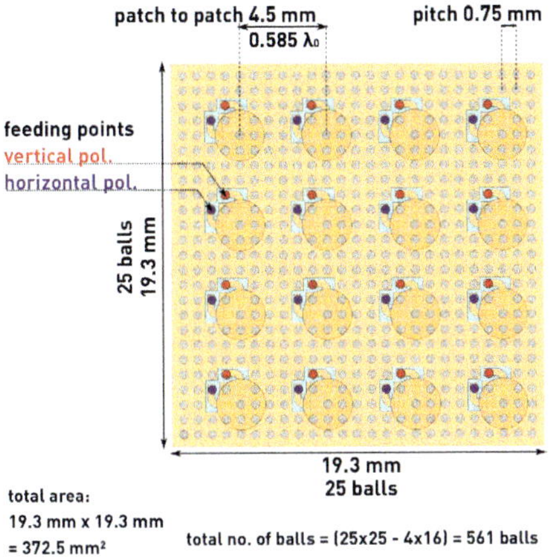

Fig. 13. The design of the BGA interface (footprint of the array module).

to electronically switch the main beam in two directions using a microstrip-line-to-slotline single-pole dual-throw (SPDT) switch. This switch utilizes commercially available PIN diodes. The antenna is fabricated using a cost-effective printed circuit board process on a CuClad 217 substrate. The antenna is small-sized and low-profile having dimensions of 12 × 28 × 10 mm (with coaxial connector).

The proposed antenna design incorporates two radiators positioned in different directions, both of which are connected to a single RF port through the SPDT switch. By altering the state of the switch, the radiation pattern of the antenna can be reconfigured. To achieve this, linear tapered slot antennas (LTSA) are used as the radiating elements. The choice of LTSA is motivated by their end-fire radiation pattern in the dielectric plane and their compatibility for integration with the PIN-diode based microstrip-to-slotline switch. The working principle of the antenna is presented in Figure 14. The antenna is fed through a 1.0 mm coaxial connector (Figure 15(a)). The resulting radiation patterns for both switching configurations (left and right) are shown in Figure 15(b). The maximum gain for both configurations is about 4 dBi.

60 GHz quadruple radiator reconfigurable antenna concept

The next energy-efficient antenna design is an alternative approach to PIN-diode based beam switching [24]. The 60 GHz antenna, presented in Figure 16, consists of eight aperture-coupled square microstrip patches arranged in a corporate feed network on the top layer. The microstrip patches are divided into four pairs, with each pair being fed through a common branch of the feeding network and a quarter-wave transformer. Both patches within each pair are excited with the same amplitude and phase. These

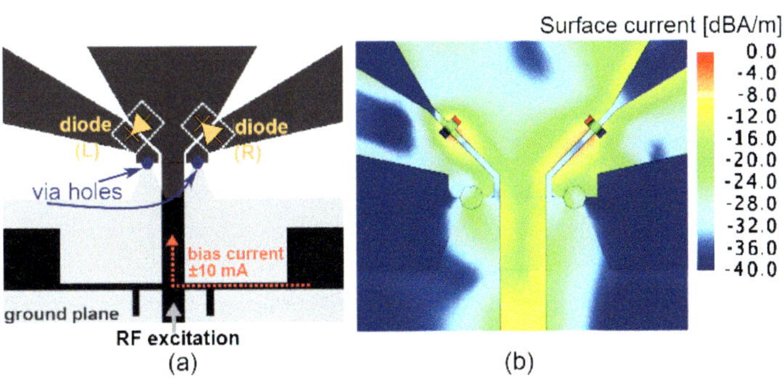

Fig. 14. The working principle of the antenna: (a) design of the reconfigurable microstrip to slotline transition, (b) surface current distribution in the R configuration – the R diode is forward biased.

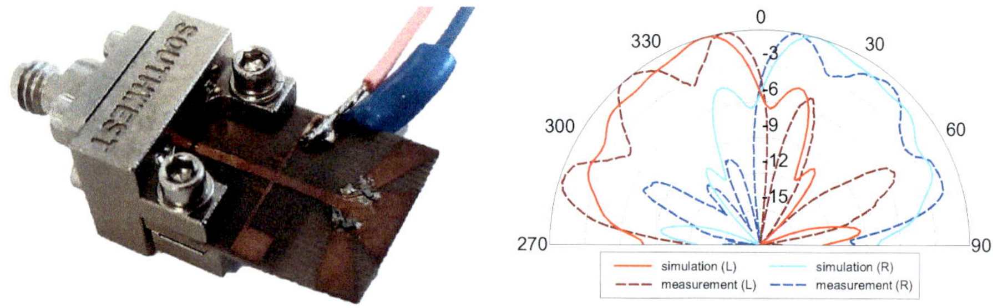

(a) (b)

Fig. 15. (a) Photograph of the manufactured antenna with soldered DC bias wires to control the switching state of the antenna. The unused solder pad is left to maintain the symmetry of the structure. (b) Comparison of simulated and measured normalized radiation patterns (dB) in the horizontal plane ($\theta = 90°$) for both switching states at the frequency equal to 60 GHz.

patches are evenly spaced along the circumference of a half-circle. The spacing between the patches is carefully chosen to minimize mutual coupling and ensure that the radiated waves from both patches within a pair are nearly in-phase. The maximum gain of a single pair is about 6 dBi.

Beam steering in the horizontal plane for the proposed antenna is achieved by selecting an active pair of microstrip patches. This is done by shorting the branches of the feed network to the antenna ground plane, effectively disabling other pairs of patches. In simulations, via-holes were employed to shorten the arms of the feeding network to the ground plane. These shorting points, located at a quarter-wavelength distance from the power splitter, appear as open circuits to the power divider. PIN diodes can be used in practical implementations for switching, as they can approximate a short or open circuit based on the applied DC

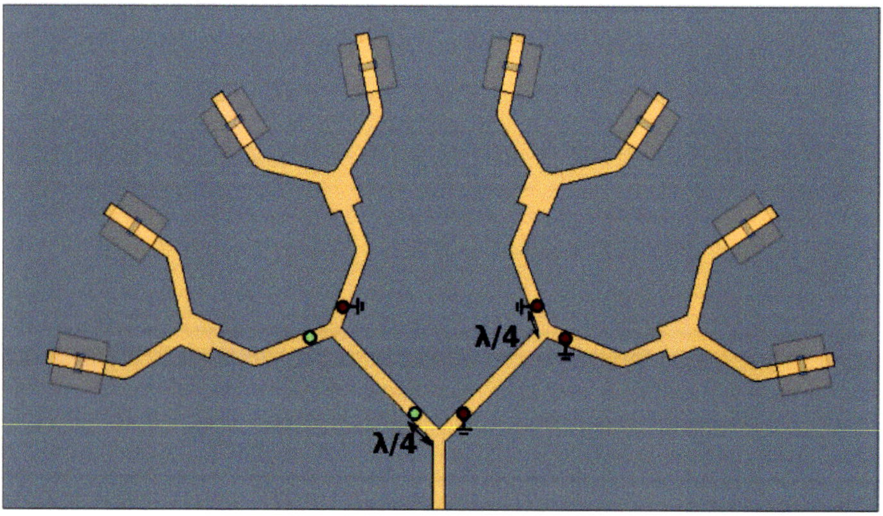

Fig. 16. Feed network steering points positions for the proposed antenna. In every considered antenna configuration, four points are shorted (marked red), while the remaining two are left open.

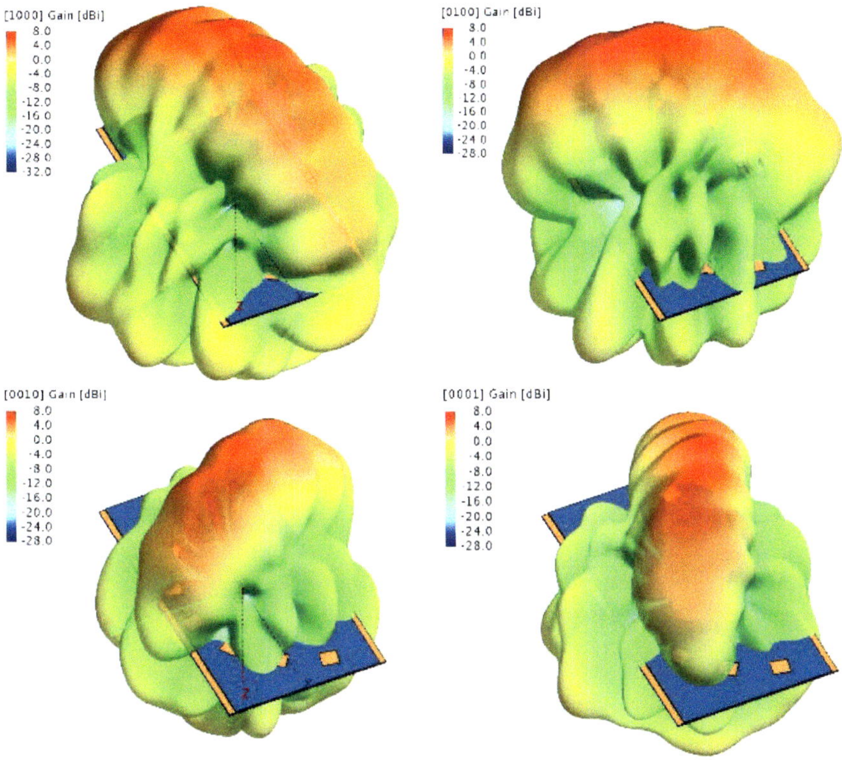

Fig. 17. Simulated 3D antenna radiation patterns.

bias which can be controlled using external digital input/output (DIO) signals. As a result, the proposed antenna having only a single RF input can provide four radiation beams (Figures 17 and 18) that can be switched using six DIO ports to control the shorting points of the feeding network. PIN diodes were chosen for the initial prototype implementation to limit potential issues of inhouse assembly. Due to issues with very thin and flexible substrate in fabricated prototype, the antenna

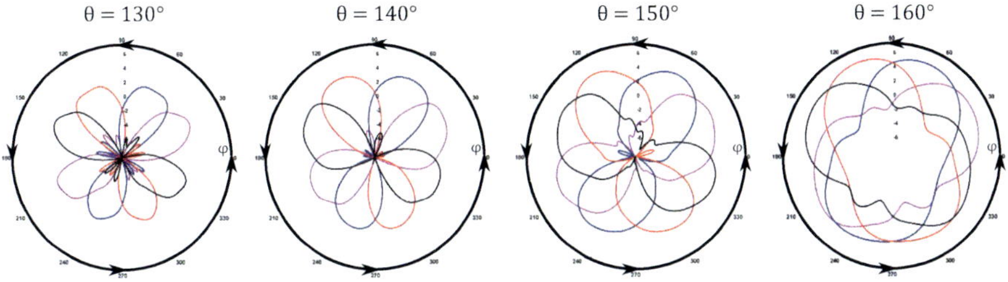

Fig. 18. 2D theta cuts of the simulated antenna radiation patterns.

has not been measured. The improved future designs will include low power consumption components, such as integrated CMOS switches, and rigid substrates that allow the realization of fully functional prototypes.

III. SUMMARY AND COMPARISON OF DIFFERENT CONSTRUCTION

All of the proposed designs were considered to cover various applications requiring unique performance of wireless communication links. Proposed techniques focus on energy-efficient beam control techniques integrated with unique antenna designs to offer connectivity, safety and security improvement with beamforming. Such solutions have limited performance in comparison to MIMO techniques but require less power to operate. Some of the considered use cases do not require fully controllable beam reshaping and simple beam modification that narrows the beam may be sufficient. An overview of each design is summarized in Table 2. Proposed designs show diverse performance, i.e. gain ranging from 3 to 15 dBi, radiation pattern from omnidirectional to narrow beam and unique radiation zones, maintaining that each of the proposed solutions may utilize single RF signal input. The multiport designs presented in sections (C) and (E) require only a simple switching circuit that concurrently would utilize single RF input from the array to offer beam shaping functionalities.

IV. CONCLUSION

In this paper, a number of various antenna designs for 5G applications have been reviewed. Simple single-input reconfigurable antennas can be used to provide not only beamforming but also direction-of-arrival estimation capabilities relying on received signal strength measurements. Therefore, the proposed approach can be applied in 5G nodes or gateways, in which low cost and power efficiency are important. Specific use cases require only a couple of discrete beams or zones from an antenna to enhance overall system performance and capabilities. Some of the proposed techniques may offer a great reduction of power consumption, while still maintaining limited beamforming capabilities in comparison to large, high-power beamforming networks that need to be attached to a large antenna array.

TABLE 2
DESIGN SUMMARY

Antenna	Frequency [GHz]	Dimensions [mm]	Gain [dBi]	3dB Beamwidth [°] E/H-plane	Unique radiation patterns
(A)	5.9	Ø110 × 22.1	3.3–5.8	Omnidirectional: 45/130	4096
(B)	5.9	Ø70 × 86.1 to Ø100 × 101.1	12.9–14.9	29.4/27.7 to 33.6/30.2	1 per lens
(C)	5.9	120 × 120 × 82	12	~28/79	4
(D)	5.9	42 × 42 × 40	7.6–8.4	58/156	5
(E)	39	55 × 41 × 26	14.3–16.9	13/23 to 18/101	32
(F)	60	12 × 28 × 10	4	80/180	2
(G)	60	25 × 27 × 10	6	50/75	4

References

[1] G. M. Rebeiz *et al.*, "Millimeter-wave large-scale phased-arrays for 5G systems," *Proc. IEEE MTT-S Int. Microw. Symp. (IMS)*, May 2015, pp. 1–3.

[2] Q. C. Li, H. Niu, A. T. Papathanassiou, and G. Wu, "5G network capacity: Key elements and technologies," *IEEE Veh. Technol. Mag.*, vol. 9, no. 1, pp. 71–78, Mar. 2014.

[3] T. S. Rappaport *et al.*, "Millimeter wave mobile communications for 5G cellular: It will work!," *IEEE Access*, vol. 1, pp. 335–349, 2013.

[4] P. Zhouyue and F. Khan, "An introduction to millimeter-wave mobile broadband systems," *IEEE Communications Magazine*, vol. 49, pp. 101-107, 2011.

[5] C. G. Christodoulou, Y. Tawk, S. A. Lane and S. R. Erwin, "Reconfigurable Antennas for Wireless and Space Applications," *Proceedings of the IEEE*, vol. 100, no. 7, pp. 2250-2261, July 2012.

[6] D.Rodrigo,B.A.CetinerandL.Jofre,"Frequency, radiation pattern and polarization reconfigurableantennausingaparasiticpixel layer," *IEEE Trans. Antennas Propag.*, vol. 62, no. 6, pp. 3422-3427, June 2014.

[7] Tang, W., Gao, S., Zhang, Y. (2020). "Lens antennas for 5G communications: Challenges and opportunities," *IEEE Communications Magazine*, 58(2), 116-122.

[8] Lin, C. H., Lee, C. H. (2021). "Energy-efficient lens antenna design for 5G communications," *IEEE Transactions on Antennas and Propagation,* 69(2), 982-993.

[9] L. Leszkowska, D. Duraj, M. Rzymowski, K. Nyka and L. Kulas, "Electronically REconfigurable Superstrate (ERES) Antenna," *2019 13th European Conference on Antennas and Propagation (EuCAP), Krakow, Poland*, 2019, pp. 1-4.

[10] Smith, D., Mysore, R. (2019). "Switched-beam antennas for energy-efficient communication in 5G networks." *IEEE Communications Magazine*, 57(3), 40-46.

[11] Chen, Z., Zheng, G., Yang, G. (2020). "Energy-efficient switched-beam antenna design for mmWave 5G communications." *IEEE Transactions on Vehicular Technology*, 69(9), 9751-9763.

[12] M. Rzymowski, P. Woznica and L. Kulas, "Single-anchor indoor localization using ESPAR antenna," *IEEE Antennas and Wireless Propagation Letters*, vol. 15, pp. 1183-1186, 2016, doi: 10.1109/LAWP.2015.2498950.

[13] Phan, A. H., Lee, C. H., Nallanathan, A. (2018). "Multibeam antenna systems for future wireless communication networks," *IEEE Communications Magazine*, 56(6), 106-112.

[14] M. Burtowy, M. Rzymowski and L. Kulas, "Low-profile ESPAR antenna for RSS-based DoA estimation in IoT applications," *IEEE Access*, vol. 7, pp. 17403-17411, 2019, doi: 10.1109/ACCESS.2019.2895740.

[15] D. Duraj, M. Rzymowski, K. Nyka and L. Kulas, "ESPAR antenna for V2X applications in 802.11p frequency band," *2019 13th European Conference on Antennas and Propagation (EuCAP), Krakow, Poland,* 2019, pp. 1-4.

[16] D. Duraj, M. Tarkowski, M. Rzymowski, L. Kulas and K. Nyka, "RSS-Based DoA Estimation in 802.11p Frequency Band Using ESPAR Antenna and PPCC-MCP Method," *2020 23rd International Microwave and Radar Conference (MIKON)*, Warsaw, Poland, 2020, pp. 152-156, doi: 10.23919/MIKON48703.2020.9253949.

[17] W. Kalista, L. Leszkowska, M. Rzymowski, K. Nyka, and L. Kulas, "Low-cost 3D printed dielectric lens antennas for 5.9 GHz frequency band V2X applications," in *Proc. 2022 24th Int. Microwave Radar Conf. (MIKON)*, Gdansk, Poland, 2022, pp. 1-4, doi: 10.23919/MIKON54314.2022.9924842.

[18] W. Kalista, L. Leszkowska, M. Rzymowski, K. Nyka and L. Kulas, "Low-cost 3D printed circularly polarized lens antenna for 5.9 GHz V2X applications," *2023 17th European Conference on Antennas and Propagation (EuCAP), Florence, Italy*, 2023, pp. 1-4, doi: 10.23919/EuCAP57121.2023.10133420.

[19] C. -Y. Lin, F. -M. Kang, P. -M. Hsu, J. -J. Chen and Y. -C. Tseng, "An accurate vehicle-to-vehicle instant alert sys-

tem using directional antennas", *2021 IEEE 94th Vehicular Technology Conference (VTC2021-Fall)*, pp. 1-5, 2021.

[20] E. Kalogeiton, D. Iapello and T. Braun, "Equipping NDN-VANETs with directional antennas for efficient content retrieval," *2020 IEEE 17th Annual Consumer Communications & Networking Conference (CCNC)*, pp. 1-8, 2020.

[21] K. Trzebiatowski, J. Fromme, D. Duraj, L. Kulas, and K. Nyka, "A dual-polarized 39 GHz 4x4 microstrip antenna array for 5G MU-MIMO airflight cabin connectivity," *Proc. 2022 24th Int. Microwave Radar Conf. (MIKON), Gdansk, Poland*, 2022, pp. 1-4, doi: 10.23919/MIKON54314.2022.9924929.

[22] K. Trzebiatowski, W. Kalista, M. Rzymowski, L. Kulas, and K. Nyka, "Multibeam antenna for Ka-band CubeSat connectivity using 3-D printed lens and antenna array," *IEEE Antennas Wireless Propag. Lett.*, vol. 21, no. 11, pp. 2244-2248, Nov. 2022, doi: 10.1109/LAWP.2022.3189073.

[23] K. Trzebiatowski, M. Rzymowski, L. Kulas, and K. Nyka, "Simple 60 GHz switched beam antenna for 5G millimeter-wave applications," in *IEEE Antennas Wireless Propag. Lett.*, vol. 20, no. 1, pp. 38-42, Jan. 2021, doi: 10.1109/LAWP.2020.3038260.

[24] M. Rzymowski, K. Trzebiatowski, K. Nyka and L. Kulas, "DoA estimation using reconfigurable antennas in millimeter-wave frequency 5G systems," *Proc. 2019 17th IEEE Int. New Circuits Syst. Conf. (NEWCAS), Munich, Germany,* 2019, pp. 1-4, doi: 10.1109/NEWCAS44328.2019.8961285.

Damian Duraj received his BSc and MSc degrees in electronics and telecommunications engineering, majoring in wireless communication engineering from Gdansk University of Technology, Poland, where he is currently a full-time researcher with the Department of Microwave and Antenna Engineering. He has conducted research and coordinating technical work in multiple national and EU projects. His research interests focus on antenna development, especially energy-efficient reconfigurable antennas along with microwave and millimeter wave circuits for improved connectivity, as well as computational electrodynamics and signal processing.

Luiza Leszkowska, born in 1994, received her BSc and MSc degrees (with honors) in electronics and telecommunications engineering, majoring in wireless communication engineering, from the Gdansk University of Technology, Poland, in 2017 and 2018, respectively. She is currently affiliated with the Department of Microwave and Antenna Engineering, Faculty of Electronics, Telecommunications and Informatics, Gdansk University of Technology. Her research interests include reconfigurable antennas and antennas for satellite applications.

Weronika Kalista, born in 1997, received her BSc and MSc degrees in electronics and telecommunications engineering, majoring in wireless communication engineering, from the Gdansk University of Technology, Poland in 2020 and 2021, respectively. She is currently employed at the Department of Microwave and Antenna Engineering, Faculty of Electronics, Telecommunications and Informatics, Gdansk University of Technology. Her research interests include dielectric lens antennas and switched beam antennas.

Kamil Trzebiatowski, born in 1995, received his BSc and MSc degrees (with honors) in electronics and telecommunications engineering, majoring in wireless communication engineering, from the Gdansk University of Technology, Poland, in 2018 and 2019, respectively. He is currently affiliated with the Department of Microwave and Antenna Engineering, Faculty of Electronics, Telecommunications and Informatics, Gdansk University of Technology, where he conducts research in the field of millimeter wave antennas, focusing on retrodirective arrays as well as reconfigurable antennas and antenna arrays.

Lukasz Kulas received his MSc and PhD degrees (with honors) in microwave engineering from Gdansk University of Technology (GUT), Gdansk, Poland, in 2001 and 2007 respectively. Currently he holds an Associate Professor position at GUT within the Department of Microwave and Antenna Engineering. His main interests are reconfigurable antennas, direction-of-arrival algorithms, dependable wireless communication, which is resistant to interferences and jamming attacks, as well as wireless embedded devices and Internet-of-Things solutions that can be applied in practical industrial applications. He actively cooperates with EU industry within a numerous R&D projects including unmanned platforms.

Krzysztof Nyka received his MSc and PhD degrees (with honors) from Gdansk University of Technology, Poland, where he is currently an Associate Professor with the Department of Microwave and Antenna Engineering. He has very actively cooperated with industry partners within many national and EU projects. His research interests include computational electrodynamics, especially the application of model order reduction in finite-element method for efficient CAD tools, as well as microwave and millimeter wave circuits and reconfigurable antennas for secure wireless applications.

Recent Progress on SOI CMOS Power Amplifiers for Mobile and WiFi Applications

Alexandre Giry, Ayssar Serhan, Ali Alshakoush, and Pascal Reynier

***Abstract*—Mobile and WiFi applications are stimulating major research efforts on next-generation power amplifiers (PAs) in order to get improved linearity and operating bandwidth with reduced power consumption. The need is for a higher integration push for the development of compact integrated PAs capable of supporting an ever-increasing number of frequency bands with high linear output power. Today, CMOS SOI technology appears as an attractive choice for efficient PA integration. This paper presents a comprehensive overview of recent CMOS SOI PA solutions targeting high-performance mobile and WiFi applications.**

***Index Terms*—CMOS, Doherty PA (DPA), efficiency, envelope tracking (ET), FOWLP, LDMOS, multimode, multiband, power amplifier (PA), reconfigurable, SOI, tunable.**

I. INTRODUCTION

THE need for mobile and WiFi RF front-ends with reduced dimensions and power consumption is driving research towards integrated PAs with challenging size and efficiency requirements [1]. Furthermore, new wireless standards such as 5G/6G and WiFi6E/7 dramatically increase output power, operating frequency, channel bandwidth and linearity requirements [2][3]. This calls for the development of integrated PA solutions capable of amplifying signals with large instantaneous bandwidth (iBW) and high peak-to-average power ratio (PAPR) while achieving high output power and linearity over an increasing frequency range. However, achieving high transmit power with high efficiency and linearity in CMOS technology is challenging given the limited power density of conventional CMOS compared to SiGe and GaAs technologies. CMOS silicon-on-insulator (SOI) appears as an attractive technology for high-power PA integration [4]. This chapter reviews recent progress related to the integration of CMOS SOI PAs for mobile and WiFi applications with emphasis on PAs designed and measured at the authors' laboratory. The chapter is organized as follows: main PA design challenges are exposed in Section 2, PA architectures with improved back-off efficiency are discussed in Section 3, an overview of recent reconfigurable PA solutions in CMOS SOI is given in Section 4, followed by a conclusion.

This work has received funding from the ECSEL Joint Undertaking (JU) in the framework of BEYOND5 (Grant agreement no. 876124) and REFERENCE (Grant agreement no. 692477) projects.

A. Giry, A. Serhan, A. Alshakoush and P. Reynier are with CEA-Leti, Grenoble, France (e-mail: {alexandre.giry, ayssar.serhan, ali.alshakoush, pascal.reynier}@cea.fr).

II. CHALLENGES

The design space of high-performance PAs is multi-dimensional and implies multiple tradeoffs between various metrics. With every new mobile and WiFi standard generation, the PA must deal with more complex modulation schemes with higher PAPR and wider iBW, together with more stringent error vector magnitude (EVM) and adjacent channel leakage ratio (ACLR) requirements, as illustrated in Figure 1.

One main PA design challenge is then to achieve high efficiency with modulated signals presenting high PAPR and wide iBW, while meeting stringent linearity requirements. To avoid signal distortion with high PAPR signals, the PA needs to be backed-off from its peak power. Since power back-off (PBO) drastically degrades PA efficiency, high-efficiency PA architectures with improved PBO efficiency need to be developed, as will be discussed in Section 3. In addition, large iBW tends to degrade PA linearity because of impedance variations and memory effects. Then, broadband linear PA architectures should be developed to support larger iBW and an increased number of operating frequency bands. Furthermore, the PA should be capable of high output power to address high power user equipment (HPUE) requirements in mobile applications. The HPUE mode is increasingly used in LTE and 5G applications to preserve cell coverage, mainly for TDD bands, by doubling the linear output power from 23 dBm (Power Class 3) to 26 dBm (Power Class 2 or HPUE) at the antenna.

CMOS technologies face serious efficiency and linearity challenges for watt-level PA applications compared to GaAs and SiGe. Indeed, active devices in modern bulk CMOS technologies are limited in terms of maximum allowed supply voltage because of low breakdown voltage and reliability concerns. Several approaches have been proposed to overcome voltage/power limitations of CMOS devices, including power combining and cascoding. These efforts allowed CMOS penetrating the PA market for 2G/3G applications, but CMOS PAs still face serious output power, linearity, and efficiency limitations for 4G/5G and WiFi-6/7 applications. Today, CMOS SOI technologies offer interesting perspectives for high-power PA integration. In [5]-[11], the stacked-FET approach [12] has been investigated for 3G/4G applications, achieving good linearity but limited output power. In [13]-[17], the use of high-voltage MOS with drain extension was considered for the design of high-efficiency multimode multiband PAs with high linear output power. LDMOS provides high thermal

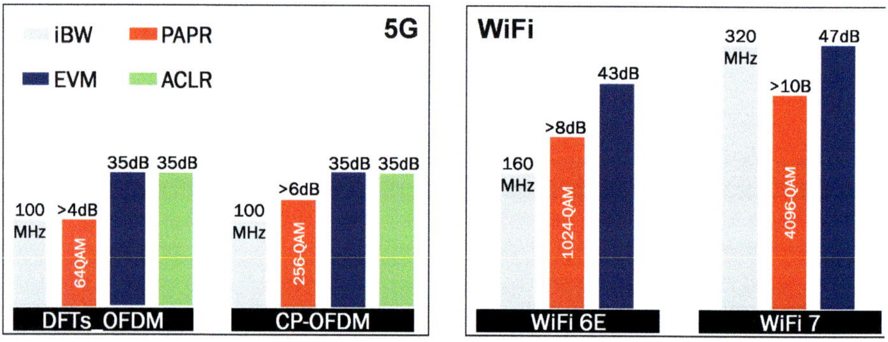

Fig. 1. Typical PA specifications for 5G and WiFi mobile applications.

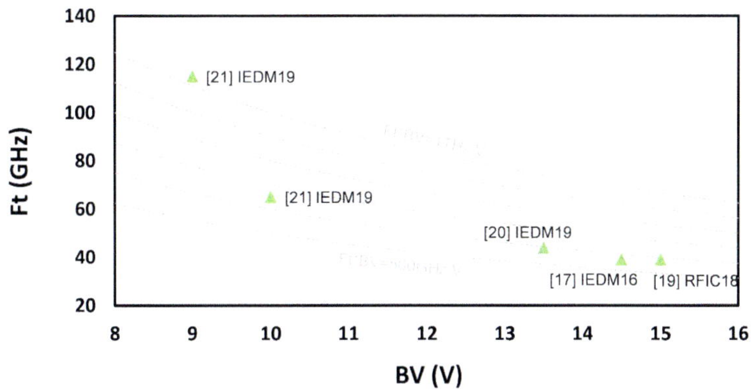

Fig. 2. Ft vs. BV of state-of-the-art LDMOS devices in SOI CMOS technologies.

stability with high ruggedness and good linearity characteristics [18] and several foundries are today offering LDMOS devices into their CMOS SOI process.

Figure 2 reports the transit frequency (Ft) and breakdown voltage (BV) of recent LDMOS devices [19]-[23] in CMOS SOI. As can be observed, their performance has been significantly improved over recent years, making it possible to consider this device for high-performance PA design.

III. HIGH-EFFICIENCY PA

As introduced in the previous section, the PA needs to achieve high efficiency not only at peak power but also at back-off power to efficiently amplify high PAPR signals. PA architectures with improved back-off efficiency generally fall into two main categories [24]: supply modulation and load modulation.

Supply Modulation

Envelope tracking (ET) is a well-known supply modulation technique [25]. During the last decade, ET has been progressively adopted in high-end handset applications to improve PA efficiency and linearity by dynamically adjusting the PA supply voltage according to the envelope of the transmitted signal [26]. To meet stringent linearity requirements, the ET modulator (ETM) must generate a modulated supply signal that is accurately aligned with the RF signal. Indeed, the delay mismatch between the supply and RF paths is a critical source of distortion [27], which becomes even more critical as the signal bandwidth increases. The most widely adopted ETM architecture combines the DC output voltage from a high efficiency DC–DC converter with a fast envelope signal from a linear amplifier which can be DC or AC-coupled [28]. 5G and WiFi systems require fast ETM with large bandwidth to accurately track the wideband envelope of the transmitted signal. Most advanced CMOS ETM can achieve up to 200 MHz of envelope signal bandwidth, as shown in Figure 3(a). In [36], a 0.18 μm BCD analog ETM chip of 1.3 mm × 1.5 mm can track a 200 MHz 5G-NR envelope signal, achieving a measured peak efficiency of 78.5% at 2.7 W peak output power. Figure 3(b) shows the efficiency characteristics of different ETMs as a function of the output power. As can be seen, ETM efficiency degrades when output power decreases. Since ETPA system efficiency is the product of ETM efficiency and PA efficiency, this degradation has a direct impact on the overall efficiency.

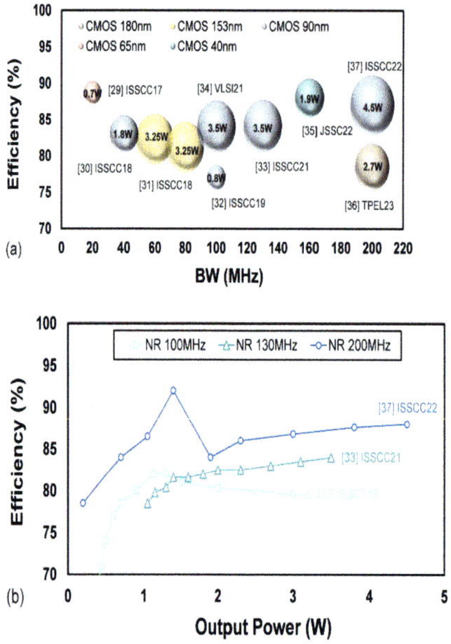

Fig. 3. State-of-the-art ETMs in CMOS technology: (a) ETM efficiency vs. bandwidth (bubble size: peak output power), (b) ETM efficiency characteristics vs. output power (extracted from [29], [30] and [33]).

In [37], a 200MHz digital ETM is proposed to overcome back-off efficiency degradation of analog ETM solutions. The digital ETM dynamically changes the PA supply voltage level by selecting one of multiple voltages from a switched-capacitor voltage divider. The proposed architecture achieves an efficiency of 87% at a peak output power of 4.5 W and a peak efficiency of 93.6% at 5 dB back-off (1.5 W output power) at the expense of higher die size (4.3 mm × 3.2 mm).

To assess practical performance of a CMOS SOI ETPA, a 2 W Class-BJ SOI LDMOS PA output stage was designed in a 130 nm RFSOI industrial process from STMicroelectronics. Simulated PAE as a function of output BO (OBO) at 2.5 GHz is reported in Figure 4, considering different ETM efficiency and shaping functions. Considering maximum efficiency shaping with an ideal ETM of 100% efficiency, the SOI ETPA can achieve higher than 50% of PAE over 13 dB OBO. Considering a real-

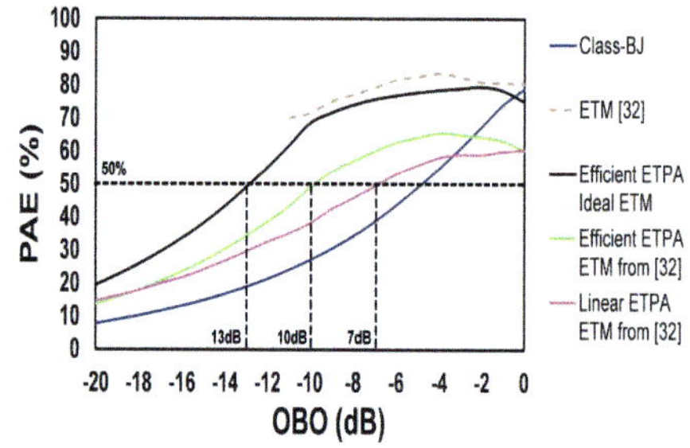

Fig. 4. SOI LDMOS ETPA output stage: simulated PAE vs. OBO (Freq = 2.5 GHz, P_{sat} = 33 dBm)

istic 100 MHz ETM [29], the SOI ETPA reaches higher than 50% of PAE over 10 dB and 7 dB OBO with maximum efficiency shaping (efficient ETPA) and constant gain shaping (linear ETPA) respectively.

Load Modulation

The Doherty PA (DPA) relies on the concept of active load modulation to maintain high efficiency from back-off to peak power [38]. It is widely adopted in high output power base station applications [39][40]. More recently, the DPA architecture was also investigated for handset applications [41], with more careful consideration of PA size and linearity due to limited space and absence of powerful digital pre-distortion (DPD). To evaluate the performance of a CMOS SOI DPA, a 2 W Class-BJ LDMOS DPA output stage was designed using the same SOI process as in the previous section. Simulated PAE as a function of OBO at 2.5 GHz is reported in Figure 5, considering differ-

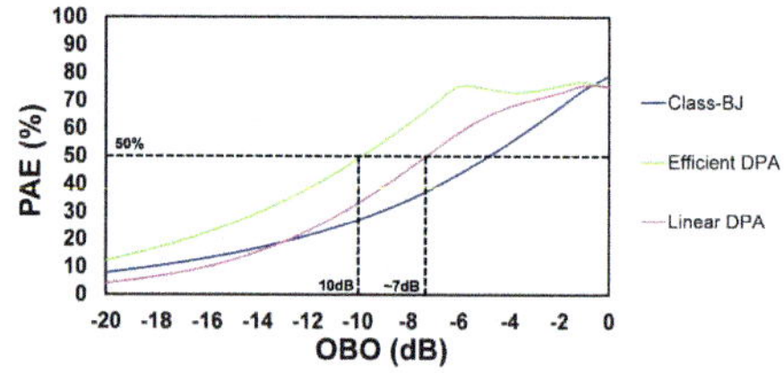

Fig. 5. SOI LDMOS DPA output stage: simulated PAE vs. OBO (Freq = 2.5 GHz, P_{sat} = 33 dBm)

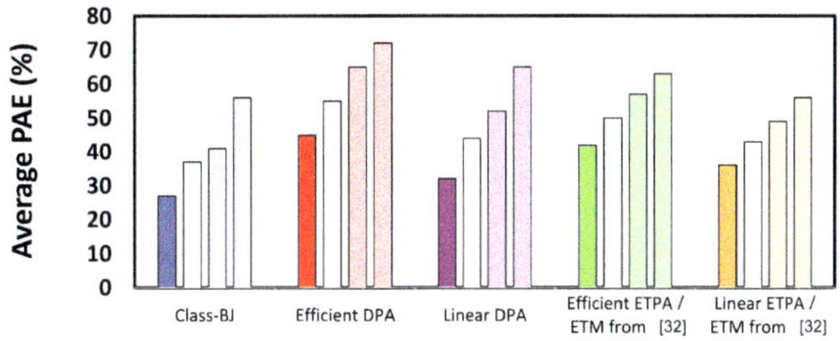

	Class-BJ	Efficient DPA	Linear DPA	Efficient ETPA / ETM from [32]	Linear ETPA / ETM from [32]
CP-OFDM 0.01%	27.00	45.00	32.00	42.00	36.00
CP-OFDM 1%	37.00	55.00	44.00	50.00	43.00
DFTs-OFDM 0.01%	41.00	65.00	52.00	57.00	49.00
DFTs-OFDM 1%	56.00	72.00	65.00	63.00	56.00

Fig. 6. Average PAE of DPA and ETPA for different modulation (DFTs-OFDM and CP-OFDM), cresting factor (0.01% and 1%) and shaping functions (efficient and linear).

ent shaping functions. The SOI DPA can reach higher than 50% of PAE over 10 dB and 7 dB OBO with maximum efficiency shaping (efficient DPA) and constant gain shaping (linear DPA) respectively.

Efficiency performance of both SOI ETPA and DPA is compared in Figure 6. For amplification of 5G DFTs-OFDM and CP-OFDM signals, the DPA architecture represents an interesting alternative solution to ETPA. Indeed, it can achieve similar efficiency without requiring complex time calibration or/and linearization nor bulky and costly ETM. In addition, by optimizing the bias and phase shift between the main and auxiliary paths, high raw linearity can be achieved, relaxing or even eliminating the need for DPD.

In [42], a two-stage LTE DPA with high efficiency and raw linearity was experimentally demonstrated. Without using DPD, the SOI DPA achieves a measured PAE and E-UTRA ACLR of 44% and –35dBc respectively at 2.5 GHz and 28 dBm average output power. The D = PA employs a lumped C-L-C inverter

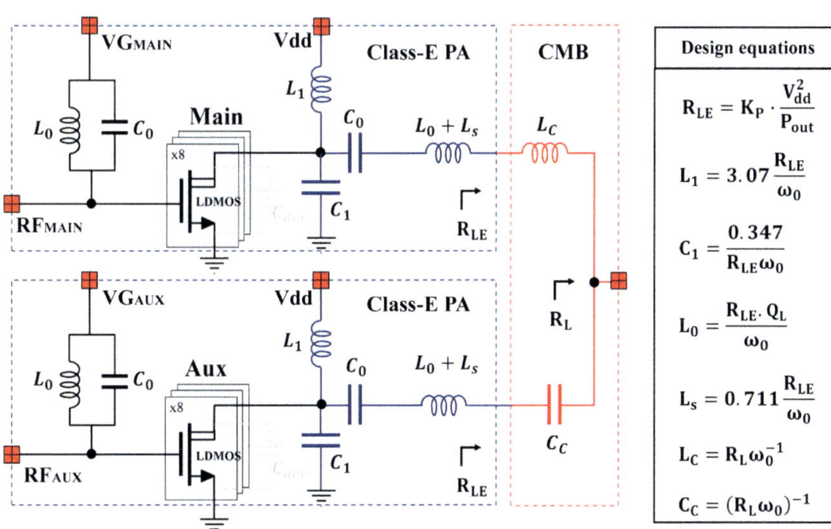

Fig. 7. Schematic and design equations of a Class-E SOI DPA with compact L-C combiner [44].

structure as output combiner. As demonstrated in [43], other DPA combiner architectures could be considered and synthetized using a set of boundary conditions at the peak and back-off points. However, the approach does not guarantee that the main transistor maintains a constant drain voltage across the BO region. Consequently, the drain efficiency may exhibit significant drop between the two efficiency peaks. To avoid this issue, a new combiner analysis method was proposed in [44], allowing ensuring maximum efficiency all over the back-off region by deriving optimal current driving profile. By applying the proposed method to a compact L-C combiner [45], it is demonstrated that up to 7.2 dB back-off can be achieved. The approach was validated on a Class-E SOI DPA using the set of equations given in Figure 7 and considering $K_p = 0.45$ and $QL = 1$.

IV. RECONFIGURABLE PA

Reconfigurable PA architectures pave the way for compact multimode multiband PA solutions by considering a single PA line-up instead of several ones. CMOS SOI reconfigurable DPA designs have been recently introduced [46]-[53].

Reconfigurable DPA for 4G/5G

DPAs generally have a limited band of operation and require precise phase control to maintain high linearity and efficiency over frequency. To overcome linearity degradation over frequency without using DPD, a 4G/5G reconfigurable DPA allowing significant linearity improvement was recently demonstrated [51]. Tunable components, integrated on the CMOS SOI die, provide improved linearity over frequency while preserving PAE and eliminating the need for DPD. The measured saturated power is higher than

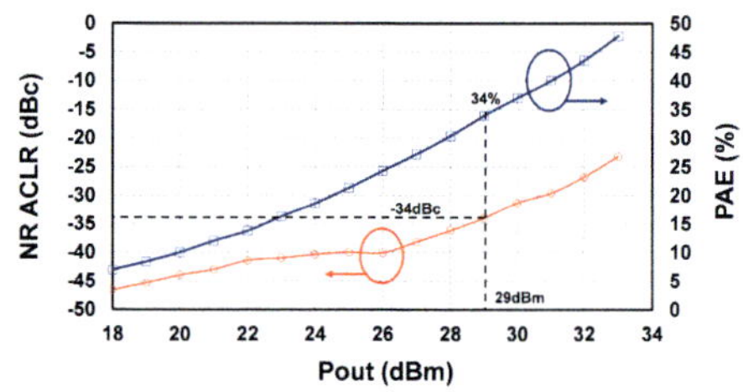

Fig. 8. NR-ACLR (left) and PAE (right) vs. P_{out} for a 5G-NR 100 MHz full-RB QPSK signal (8dB PAPR) at 2.3 GHz and $V_{dd} = 5$ V.

32 dBm over 1.9 GHz to 2.7 GHz. The corresponding PAE at peak power and at 4.5 dB BO is higher than 50% and 40% respectively from 2 GHz to 2.6 GHz. With an LTE signal, the proposed PA achieves 43.5% of PAE and an E-UTRA ACLR of –39.6 dBc at 28 dBm. In HPUE mode, the PA reaches a saturated power of 36 dBm with a maximum PAE of 57% and delivers a linear P_{out} of 31 dBm with 42.6% of PAE and –35.7 dBc of ACLR. The PA also supports 5G, as shown in Figure 8 with a 100 MHz 5G-NR signal.

Reconfigurable DPA for WiFi

High-power Wi-Fi 6/6E RF front-ends face significant challenges due to stringent linearity requirements. The PA must deliver more than 20 dBm of average output power with better than –43 dB of EVM under 1024-QAM OFDM signal. In [52], a robust high-power Wi-Fi 6 PA module composed of an SOI-CMOS PA core embedded in a low-cost two-layer fan-out wafer-level package (FOWLP) was demonstrated. The SOI PA die features a two-stage linear DPA that includes all the active and passive components, except two inductors of the output matching network that are integrated into the FOWLP. The DPA achieves 35.1 dBm of P_{sat} with 53% of peak PAE and state-of-the-art average P_{out} and EVM levels under Wi-Fi

6/6E signals under 5 V supply. With an 802.11ac MCS11-1024QAM 40MHz signal the PA delivers a linear P_{out} of 22.5 dBm while achieving an EVM of –43 dB without DPD. The PA shows robust operation even under extreme load mismatch, input power and temperature conditions. In [53], a high-power monolithic CMOS SOI Wi-Fi 6/6E front-end module composed of a multimode TX path and multiple RX paths was demonstrated. The SOI die includes a reconfigurable DPA, a low noise amplifier, an antenna switch, and digital control. The RX path achieves 15 dB of gain, less than 1.8 dB of NF, –8.5 dBm of IP1dB and 3.5 dBm of IIP3. The TX path achieves 2 W of saturated power and state-of-the-art linear P_{out} and EVM levels under 3.85 V supply. The TX path can be reconfigured to provide optimum linearity and efficiency as a function of the transmitted MCS signal. Without DPD, it delivers a measured linear P_{out} of 25/23.6/20/19 dBm while achieving an EVM of –32/–34.9/–43.9/–47.2dB with a MCS7/MCS9/MCS11/MCS13 signal respectively. Figure 9 shows the measured EVM as a function of output power achieved without DPD with a 4096QAM MCS13 signal.

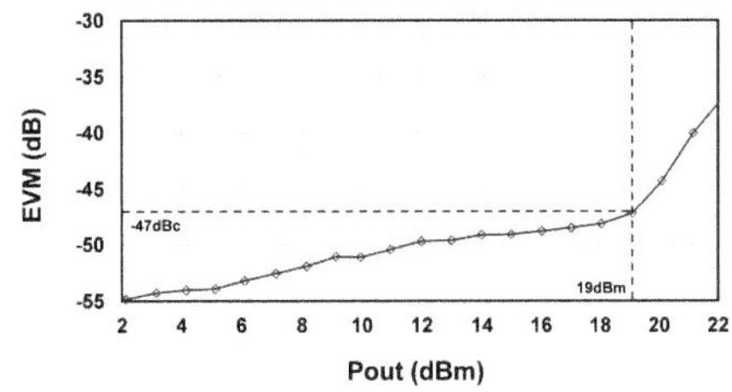

Fig. 9. EVM vs. output power at 2.4 GHz and V_{dd} = 3.85 V for a MCS13 BE40 WiFi signal.

TABLE 1
STATE-OF-THE-ART 5G-FR1 PA. [5G SIGNAL: DFTS-OFDM QPSK 100 MHZ)]

Ref	Freq [GHz]	VDD [v]	Pout [dBm]	PAE [%]	ACLR [dBc]	DPD	Process	Architecture
[51] T.W.	**2.3**	**5**	**29**	**34**	**–34**	**no**	**SOI**	**Doherty**
[56]	3.6	3.4	29.6	32	–35	yes	GaAs	AC-Stack
[55]	2.45	5.5	30.5	35.5	–35	yes	GaAs	Doherty
[59]	3.75	3.8	27.8	42	–38.8	yes	GaAs	Doherty
[54]	3.55	4.5	27.1	30	–33	no	GaAs	Doherty
[58]	3.75	3.8	27.4	39	–38	yes	GaAs	Doherty
[57]	3.75	5	31.7	40.7	–37.7	yes	GaAs	LMBA

TABLE 2
STATE-OF-THE-ART 2.4 GHZ WIFI PA

Ref	Freq [GHZ]	VDD [V]	Mod.	BW [MHz]	Pout [dBm]	EVM [dB]	Process	Architecture
[53] T.W.	**2.4**	**3.85**	**256QAM 1024QAM 4096QAM**	**40**	**23.4 20 19**	**–35.1 –43.9 –47.6**	**SOI**	**Doherty**
[60]	2.45	3.3	256QAM	20	19.2	–32	CMOS	Differential
[61]	2.4	3.3	256QAM	NA	19.9	–35	SiGe	Flip-chip

As shown in the following two tables, the proposed CMOS SOI reconfigurable DPAs provide state-of-the-art performance when compared to other PAs in SiGe and GaAs technologies operating at similar operating frequencies.

V. CONCLUSION

In this chapter, we provided an overview of recent CMOS SOI PA solutions for mobile and WiFi applications. The use of the LDMOS device, available in most SOI-CMOS processes, is crucial for reliable high-power/voltage operation and it brings serious advantages in terms of output power and efficiency. It was shown that the classical bandwidth limitation of the DPA can be overcome by efficient integration of reconfigurable SOI DPA solutions, paving the way to compact multimode multiband PAs with high-performance. Despite the substantial progress made, there are still areas for further research. In particular, the development of high-power CMOS SOI PAs that can operate at higher frequencies (>5 GHz) is crucial to meet the growing demand for higher data rates and lower latency communications. As researchers continue to focus on exploring innovative approaches, the opportunities for CMOS SOI PAs will continue to expand and evolve.

REFERENCES

[1] S. Tanaka *et al.*, "evolution of power amplifiers for mobile phone terminals from the 2nd generation to the 5th generation," *IEICE Trans. Electron.*, vol. E105.C, no. 10, pp. 421–432, Oct. 2022.

[2] T. Qi and S. He, "Power up potential power amplifier technologies for 5G applications," *IEEE Microwave*, vol. 20, no. 6, pp. 89–101, Jun. 2019.

[3] L. Ward, J. Köpp, "IEEE 802.11be Technology Introduction", White paper, Rohde&Schwarz, 2022.

[4] A. Giry *et al.*, "Linear power amplifiers for sub-6GHz mobile applications: progress and trends," *18th IEEE International New Circuits and Systems Conference (NEWCAS), Montréal, QC, Canada, Jun. 2020.*

[5] S. Pornpromlikit *et al.*, "A watt-level stacked-FET linear power amplifier in silicon-on-insulator CMOS," *IEEE Trans. Microwave Theory Techn.*, vol. 58, no. 1, pp. 57–64, Jan. 2010.

[6] M. -S. Jeon, J. Woo, U. Kim, and Y. Kwon, "High-efficiency CMOS stacked-FET power amplifier for W-CDMA applications using SOI technology," *Electron. Lett.*, vol. 49, no. 8, pp. 564–566, Apr. 2013.

[7] U. Kim and Y. Kwon, "A high-efficiency SOI CMOS stacked-FET power amplifier using phase-based linearization," *IEEE Microw. Wireless Compon. Lett.*, vol. 24, no. 12, pp. 875–877, Dec. 2014.

[8] B. Francois and P. Reynaert, "Highly linear fully integrated wideband RF PA for LTE-advanced in 180-nm SOI," *IEEE Trans. Microwave Theory Techn.*, vol. 63, no. 2, pp. 649–658, Feb. 2015.

[9] S. Park, J.-L. Woo, U. Kim, and Y. Kwon, "Broadband CMOS stacked RF power amplifier using reconfigurable interstage network for wideband envelope tracking," *IEEE Trans. Microwave Theory Techn.*, vol. 63, no. 4, pp. 1174–1185, Apr. 2015.

[10] K. Kim, D.-H. Lee, and S. Hong, "A quasi-Doherty SOI CMOS power amplifier with folded combining transformer," *IEEE Trans. Microwave Theory Techn.*, vol. 64, no. 8, pp. 2605–2614, Aug. 2016, doi: 10.1109/TMTT.2016.2577584.

[11] M. Khorshidian and H. Krishnaswamy, "A fully-integrated 2.6GHz stacked switching power amplifier in 45nm SOI CMOS with >2W output power and 43.5% efficiency," *IEEE MTT-S International Microwave Symposium (IMS), Boston, MA, USA, Jun. 2019.*

[12] P. Asbeck, "Stacked Si MOSFET strategies for microwave and Mm-wave power amplifiers," *IEEE 14th Topical Meeting on Silicon Monolithic Integrated*

Circuits in RF Systems, Newport Beach, CA, USA, 2014.

[13] A. Giry *et al.*, "A monolithic watt-level SOI LDMOS linear power amplifier with through silicon via for 4G cellular applications," *2013 IEEE Topical Conference on Power Amplifiers for Wireless and Radio Applications, Austin, TX, USA*, Jan. 2013, pp. 19–21.

[14] K. Kim et al., "A dual-mode multi-band second harmonic controlled SOI LDMOS power amplifier," *IEEE Microw. Wireless Compon. Lett.*, vol. 25, no. 7, pp. 466–468, Jul. 2015.

[15] G. Jeong, T. Joo, and S. Hong, "Highly linear SOI LDMOS power amplifier with combinated cascode structure," *2016 IEEE International Symposium on Radio-Frequency Integration Technology (RFIT), Taipei, Taiwan*, Aug. 2016.

[16] Z. Zhao et al., "A 2.3-2.7 GHz power amplifier with piecewise adaptive bias and reactance compensated interstage matching network in 130nm SOI LDMOS process," *7th International Conference on Integrated Circuits and Microsystems (ICICM), Xi'an, China, Oct. 2022*, pp. 345–348.

[17] V. Knopik et al., "A 1.95GHz 28dBm fully integrated packaged power amplifier presenting a 3G FOM of 80 (PAE+ACLR) designed in H9SOIFEM CMOS 130nm: Development of an optimized high performances RF SOI power cell," *IEEE SOI-3D-Subthreshold Microelectronics Technology Unified Conference (S3S), Burlingame, CA, USA, Oct. 2016.*

[18] T. Johansson and J. Fritzin, "A review of watt-level CMOS RF power amplifiers," *IEEE Transactions on Microwave Theory and Techniques*, vol. 62, no. 1, pp. 111-124, Jan. 2014.

[19] R. T. Toh *et al.*, "A 300mm foundry HR-SOI technology with variable silicon thickness for integrated FEM applications," *IEEE International Electron Devices Meeting (IEDM), San Francisco, CA, USA, Dec. 2016.*

[20] R. T. Toh *et al.*, "RF performance of a highly linear power amplifier EDNMOS transistor on trap-rich SOI," *IEEE Electron Device Lett.*, vol. 39, no. 9, pp. 1346–1349, Sep. 2018.

[21] R. T. Toh *et al.*, "A CMOS-SOI power amplifier technology using EDNMOS for sub 6 GHz wireless applications," *IEEE Radio Frequency Integrated Circuits Symposium (RFIC), Philadelphia, Jun. 2018.*

[22] X. Garros *et al.*, "A very robust and reliable 2.7GHz +31dBm Si RFSOI transistor for power amplifier solutions," *IEEE International Electron Devices Meeting (IEDM), San Francisco, CA, USA, Dec. 2019.*

[23] T. V. Dinh *et al.*, "Record high-performance RF devices in an advanced FDSOI process enabling integrated Watt-level power amplifiers for WiFi and 5G applications," *IEEE International Electron Devices Meeting (IEDM), San Francisco, CA, USA, Dec. 2019.*

[24] G. T. Watkins, "The best of both worlds: The dynamic load-modulation power amplifier," *IEEE Microwave*, vol. 21, no. 4, pp. 76–86, Apr. 2020.

[25] P. Asbeck and Z. Popovic, "ET comes of age: Envelope tracking for higher-efficiency power amplifiers," *IEEE Microwave*, vol. 17, no. 3, pp. 16–25, Mar. 2016.

[26] K. Mukai *et al.*, "Recent progress in envelope tracking power amplifiers for mobile handset systems," *IEICE Trans. Electron.*, vol. E104.C, no. 10, pp. 516–525, Oct. 2021.

[27] J. Jeong *et al.*, "Time mismatch effect in linearity of hybrid envelope tracking power amplifier," *IEEE Microwave and Wireless Components Letters*, vol. 25, no. 8, pp. 550-552, Aug. 2015.

[28] S. Bhardwaj *et al.*, "A review of hybrid supply modulators in CMOS technologies for envelope tracking PAs," *IEEE Trans. Power Electron.*, vol. 38, no. 5, pp. 6036–6062, May 2023.

[29] X. Liu et al., "2.4 A 2.4V 23.9dBm 35.7%-PAE -32.1dBc-ACLR LTE-20MHz envelope-shaping-and-tracking system with a multiloop-controlled AC-coupling supply modulator and a mode-switching PA," *IEEE International*

Solid-State Circuits Conference (ISSCC), San Francisco, CA, USA, Feb. 2017.

[30] C.-Y. Ho et al., "An 87.1% efficiency RF-PA envelope-tracking modulator for 80MHz LTE-Advanced transmitter and 31dBm PA output power for HPUE in 0.153μm CMOS," *IEEE Int. Solid-State Circ. Conf. (ISSCC), Feb. 2018.*

[31] T. Nomiyama et al., "A 2TX supply modulator for envelope-tracking power amplifier supporting intra- and inter-band uplink carrier aggregation and power class-2 high-power user equipment," *IEEE International Solid - State Circuits Conference (ISSCC), San Francisco, CA, Feb. 2018.*

[32] J.-S. Paek et al., "An 88%-efficiency supply modulator achieving 1.08μs/V fast transition and 100MHz envelope-tracking bandwidth for 5G new radio RF power amplifier," *IEEE International Solid-State Circuits Conference (ISSCC), San Francisco, CA, USA, Feb. 2019.*

[33] D. Kim et al., "A hybrid switching supply modulator achieving 130MHz envelope-tracking bandwidth and 10W output power for 2G/3G/LTE/NR RF power amplifiers," *IEEE International Solid- State Circuits Conference (ISSCC), San Francisco, CA, USA, Feb. 2021.*

[34] J.-S. Paek et al., "A 5G new radio SAW-less RF transmitter with a 100MHz envelope tracking HPUE n77 power amplifier module," *Symposium on VLSI Circuits, Kyoto, Japan: IEEE, Jun. 2021.*

[35] W. Leng et al., "Envelope tracking supply modulator with trellis-search-based switching and 160-MHz capability," *IEEE J. Solid-State Circuits*, vol. 57, no. 3, pp. 719–733, Mar. 2022.

[36] P. Xu et al., "A 2.7 W AC-coupled hybrid supply modulator achieving 200 MHz envelope-tracking bandwidth for 5G new radio power amplifier," *IEEE Trans. Power Electron.*, pp. 7416–7427, Jun. 2023.

[37] J.-S. Bang et al., "2-Tx digital envelope-tracking supply modulator achieving 200MHz channel bandwidth and 93.6% efficiency for 2G/3G/LTE/NR RF power amplifiers," *IEEE International Solid- State Circuits Conference (ISSCC), San Francisco, CA, USA, Feb. 2022.*

[38] W. H. Doherty, "A new high efficiency power amplifier for modulated waves," *Proc. IRE*, vol. 24, no. 9, pp. 1163–1182, Sep. 1936.

[39] V. Camarchia, M. Pirola, R. Quaglia, S. Jee, Y. Cho, and B. Kim, "The Doherty power amplifier: Review of recent solutions and trends," *IEEE Trans. Microwave Theory Techn.*, Feb. 2015.

[40] R. Pengelly, C. Fager, and M. Ozen, "Doherty's legacy: A history of the Doherty power amplifier from 1936 to the present day," *IEEE Microwave*, Feb. 2016.

[41] Y. Cho, D. Kang, K. Moon, D. Jeong, and B. Kim, "A handy dandy Doherty PA: A linear doherty power amplifier for mobile handset application," *IEEE Microwave*, Sep. 2017.

[42] A. Serhan et al., "A broadband high-efficiency SOI-CMOS PA module for LTE/ LTE-A handset applications," *IEEE Radio Frequency Integrated Circuits Symposium (RFIC), Boston, MA, USA, Jun. 2019.*

[43] M. Özen, K. Andersson and C. Fager, "Symmetrical Doherty power amplifier with extended efficiency range," *IEEE Transactions on Microwave Theory and Techniques*, vol. 64, no. 4, pp. 1273-1284, 2016.

[44] J. Bachi et al., "A novel approach for Doherty PA design using a compact L-C combiner," *IEEE Trans. Circuits Syst.* II, Oct. 2022.

[45] K. Takenaka et al., "New compact Doherty power amplifier design for handset applications," *IEEE Topical Conference on RF/Microwave Power Amplifiers for Radio and Wireless Applications (PAWR), Phoenix, AZ, USA, Jan. 2017.*

[46] U. Kim et al., "A multiband reconfigurable power amplifier for UMTS handset applications," *IEEE Trans. Microwave Theory Techn.*, vol. 60, no. 8, pp. 2532–2542, Aug. 2012.

[47] S. Kang et al., "A multi-mode multi-band reconfigurable power amplifier for

2G/3G/4G handset applications," *IEEE Microw. Wireless Compon. Lett.*, vol. 25, no. 1, pp. 49–51, Jan. 2015.

[48] U. Kim et al., "A single-chain multiband reconfigurable linear power amplifier in SOI CMOS," *IEEE MTT-S International Microwave Symposium, Phoenix, AZ, USA, May 2015.*

[49] P. Ferris et al., "A 130-nm SOI CMOS reconfigurable multimode multiband power amplifier for 2G/3G/4G handset applications," *IEEE Radio Frequency Integrated Circuits Symposium (RFIC), San Francisco, CA, USA, May 2016.*

[50] P. Draxler and J. Hur, "A multi-band CMOS Doherty PA with tunable matching network," *IEEE MTT-S International Microwave Symposium (IMS), Honolulu, HI, USA, Jun. 2017.*

[51] A. Serhan et al., "A Reconfigurable SOI CMOS Doherty power amplifier module for broadband LTE high-power user equipment applications," in *IEEE Radio Frequency Integrated Circuits Symposium (RFIC), Los Angeles, CA, USA, Aug. 2020.*

[52] P. Reynier et al., "A high-power SOI-CMOS PA module with fan-out wafer-level packaging for 2.4 GHz Wi-Fi 6 applications," *IEEE Radio Frequency Integrated Circuits Symposium (RFIC), Atlanta, GA, USA, Jun. 2021.*

[53] D. Parat et al., "A linear high-power reconfigurable SOI-CMOS front-end module for WI-FI 6/6E applications," *IEEE Radio Frequency Integrated Circuits Symposium (RFIC), Denver, CO, USA, Jun. 2022.*

[54] H. Oh *et al.*, "2.8-3.8 GHz broadband InGaP/GaAs HBT Doherty power amplifier IC for 5G new radio handset," *IEEE/MTT-S International Microwave Symposium - IMS 2022, Denver, CO, USA, Jun. 2022.*

[55] S. Imai, H. Okabe, and S. Tanaka, "A bandwidth-optimized transformer-based Doherty power amplifier for 5G Power Class 2 Handset operation at 2.2 GHz-2.7 GHz," *IEEE MTT-S International Microwave Symposium (IMS), Atlanta, GA, USA, Jun. 2021.*

[56] J. Enomoto *et al.*, "AC-stacked power amplifier for 5G mobile handset applications in Band n77," *IEEE International Symposium on Radio-Frequency Integration Technology (RFIT), Hiroshima, Japan, Sep. 2020.*

[57] K. Takenaka, Y. Noguchi, S. Arayashiki, and T. Wada, "Load-modulated balanced amplifier design for handset applications," *IEEE Microw. Wireless Tech. Lett.*, vol. 33, no. 6, pp. 855–858, Jun. 2023.

[58] K. Takenaka, Y. Noguchi, S. Arayashiki, and T. Wada, "Multiband Doherty power amplifier design for 5G NR sub-6 GHz handset applications," *Asia-Pacific Microwave Conference (APMC), Yokohama, Japan, Nov. 2022.*

[59] K. Takenaka, Y. Noguchi, Y. Takenouchi, H. Okabe, and T. Wada, "Parallel plate coupler based Doherty power amplifier design for 5G NR handset applications," *IEEE MTT-S International Microwave Symposium (IMS), Atlanta, GA, USA, Jun. 2021.*

[60] H. Ahn *et al.*, "A fully integrated −32-dB EVM broadband 802.11abgn/ac PA with an external PA driver in WLP 40-nm CMOS," *IEEE Trans. Microwave Theory Techn.*, vol. 67, no. 5, pp. 1870–1882, May 2019.

[61] A. Samelis *et al.*, "A fully integrated flip-chip SiGe BiCMOS power amplifier for 802.11ac applications," *IEEE Radio Frequency Integrated Circuits Symposium (RFIC), San Francisco, CA, USA, May 2016.*

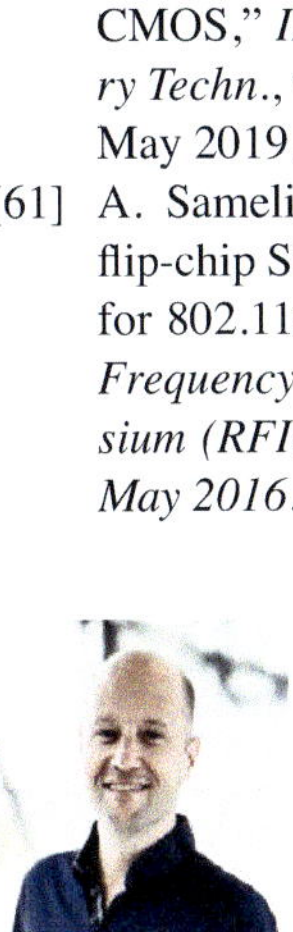

Alexandre Giry received his Ph.D. degree in Electrical Engineering from Université Grenoble Alpes (UGA), France. He is currently with CEA-Leti as a research scientist, leading research activities on RF power IC/MMIC design in silicon and GaN technologies. His current research interests include PA design in CMOS/BiCMOS/GaN/InP technologies for RF/mmW applications, high-efficiency multimode multiband PA architectures, reconfigurable and mismatch-resilient RF front-end architectures, power management for high-efficiency

PA, packaging solutions for integrated RF/mmW modules.

Ayssar Serhan received his M.Sc. degree and his Ph.D. degree in nanoelectronics and nanotechnologies from the University of Grenoble Alpes (UGA), France in 2011 and 2015, respectively. In 2015, he joined the CEA-Leti as a RF/mmWave front-end module design engineer. His research interests include the design of advanced RF PA/FEM architectures (Doherty, LMBA, outphasing, etc.) for WIFI, LTE and 5G applications; the design-for-reliability of GaN, CMOS and SOI power amplifiers; the design of tunable RF circuits including power amplifiers and antenna tuners in SOI technologies. He has a strong experience in design and simulation flow development for highly integrated multi-technology RF modules.

Ali Alshakoush obtained his M.Sc. degree in wireless communications and a Ph.D. in nanoelectronics and nanotechnologies from the University of Grenoble Alpes (UGA), France in 2017 and 2021, respectively. He joined CEA-Leti, Grenoble in 2021 as a research engineer in the field of low power RF transceivers. He is currently involved in the implementation and characterization of high-efficiency RF front-ends.

Pascal Reynier received his Ph.D. degree in electrical engineering from INSA Lyon in collaboration with STMicroelectronics in 2009. From 2008 to 2017 he was with ACCO Semi, France working on cellular RF PA and FEM products design in CMOS technology. He is currently with CEA-Leti Minatec, Grenoble, France, working as an RF/mmW IC & Package designer. Pascal has a strong background in RF PA and module/package design, with several years of hand-on experience in PA/FEM products and successful collaborative projects with industrial and academic partners. He has also a strong experience in design flows for multi-technology RF/mmW modules. His current research interests include the design and characterization of high efficiency PA/FEM solutions in silicon and III–V technologies for RF/mmW applications.

V2X RF Front End Module Assistance Circuits: Observation Receiver, DPD, and Supply Modulator

Emre Ulusoy, Bahadır Özkan, Fatih Maden, Furkan Barin, Ercem Yeşil, Adem Eren, Dursun Baran, Tufan Coşkun Karalar, Ahmet Tekin, and Ertan Zencir

***Abstract*—In today's connectivity ecosystem, the connection of not only mobile devices but also pedestrians, traffic infrastructures, ground stations and autonomous vehicles, with each other and with the environment, is growing. In this proposed example work, an observation receiver operating in the n47 frequency band (5855–5925 MHz), a digital pre-distorter (DPD), and a 40 MHz bandwidth supply modulator are presented for vehicle-to-everything (V2X) connectivity. These assistant blocks are located on the same die as the RF power amplifier. According to the simulations, the observation receiver has a maximum bandwidth of 120 MHz, which can implement DPD in systems using 5G NR signals with a bandwidth of 10/20/30/40 MHz. The low-IF architecture receiver transmits the very high swing signal from the PA output to the data converter with the help of a programmable attenuator. LUT-based DPD is a hard-coded IP with third-order memory. The supply modulator, which can provide 2 A current up to 5 V output voltage, has over 85% efficiency for 5G NR signals with 8 dB PAPR.**

***Index Terms*—Vehicle-to-everything (V2X), digital pre-distortion (DPD), integrated DPD, supply modulator.**

This paper is a result of the BEYOND5 project, which has received funding from the ECSEL Joint Undertaking (JU) under grant agreement No. 876124. The JU receives support from the European Union's Horizon 2020 research and innovation programme and France, Germany, Turkey, Sweden, Belgium, Poland, Netherlands, Israel, Switzerland, and Romania.

E. Ulusoy, B. Özkan, F. Maden, F. Barin, A. Eren, D. Baran, and E. Zencir are with TÜBİTAK, Informatics and Information Technologies Research Center (BİLGEM), Kocaeli, Turkey. E. Yeşil is with Department of Electrical and Computer Eng., University of California Los Angeles, California, United States. A Eren is with Department of Electronics Engineering, Gebze Technical University, Kocaeli, Turkey. E. Ulusoy, B. Özkan, F. Maden, F. Barin, E. Yeşil, M. Çoban, and T. C. Karalar are also with Department of Electronics and Communication Engineering, Istanbul Technical University, Istanbul, Turkey. A. Tekin is with Department of Electrical and Electronics Engineering, Boğaziçi University, Istanbul, Turkey. E. Zencir is also with Department of Electrical and Electronics Eng., University of Turkish Aeronautical Association, Ankara, Turkey. The contact email for this publication is emre.ulusoy@tubitak.gov.tr.

I. INTRODUCTION

CELLULAR communication systems, such as C-V2X, 4G, 5G, and beyond generations, are considered the smartest electronic systems globally. These systems are subject to various performance requirements defined by standards. The complexity of these requirements depends on factors such as modulation type, carrier frequency, and the utilization of multiple-input-multiple-output (MIMO) techniques in the applications. In order to facilitate "intelligent transportation systems (ITS)" applications, the European Telecommunications Standards Institute (ETSI) has allocated the 70 MHz band within the 5855–5925 MHz range. Furthermore, according to the 3GPP as of Release 16, the designated frequency band for C-V2X systems is 5855–5925 MHz (n47). The n47 band supports modulations up to 64QAM with 10, 20, 30, and 40 MHz bandwidth options.

This chapter examines auxiliary blocks through an example linearity and efficiency improved RF front-end SoC (system-on-chip) that uses CMOS SOI (silicon-on-insulator) technology. Within the example SoC, a system that can be used in the vehicle-to-everything (V2X) application and fulfills the performance and functionality requirements of the relevant standard has been designed. The example SoC was sent to production and is expected to be delivered from the foundry. It is hoped that the power amplifier in the front-end will have the performance to meet the necessary requirements without the need for III–V group materials. In order to meet these requirements, some auxiliary blocks have been used in order to achieve better linearity levels due to the material type of processes such as GaAs, GaN. We aim to increase the linearity of the CMOS power amplifiers with the methods available in the literature as digital pre-distortion techniques. A core FEM design consisting of RF PA, LNA and a high-power RF switch is the work of another research group and the details are not available in this chapter.

Another essential parameter of power amplifiers is the energy efficiency. Energy efficiency not only affects the end user positively in terms of cost (generator or battery), but also makes a small contribution to energy savings on a global scale. A supply modulator, another auxiliary block of the power amplifier, modulates the supply voltage according to the envelope of the signal at the PA input. Thus, it increases the energy efficiency.

II. ARCHITECTURE OF AN EXAMPLE RF FRONT-END SYSTEM

The architectural block diagram of the example system is given in Figure 1. It includes the core FEM blocks such as RF power amplifier, LNA and RF switch, as well as the observation receiver, digital pre-distortion (DPD) and supply modulator, which are assistance circuits. The first stage of the observation receiver is an attenuator with approximately 1 dB steps. The sensitivity of the steps of this attenuator is optimally determined; taking into account the design difficulties, the sensitivity closest to this range is provided. There is a down-converter mixer circuit following the attenuator. Since the linearity of the observation receiver is very important in DPD systems, a passive mixer and a transimpedance amplifier are used to down-convert the RF signal to IF frequency. The frequency synthesizer block that provides the LO signal to the mixer is a PLL and operates in the 5.9 GHz band. The mixer output is sampled at the ADC input through a baseband filter with adjustable gain and bandwidth.

In the low-IF architecture, the IF frequency is chosen around 60 MHz. For the ADC sampling rate specified to be around 240 msps, the IF frequency corresponding to Fs/4 is chosen because it provides an extra convenience for the digital design. Traditionally, ADC output bins are subjected to digital-down-conversion (DDC) by multiplying them by the digital LO signal generated by direct-digital-synthesis (DDS). However, for a frequency of 240 MHz in digital signal processing, the performance of the digital design in the 280 nm process has been approached with suspicion. For

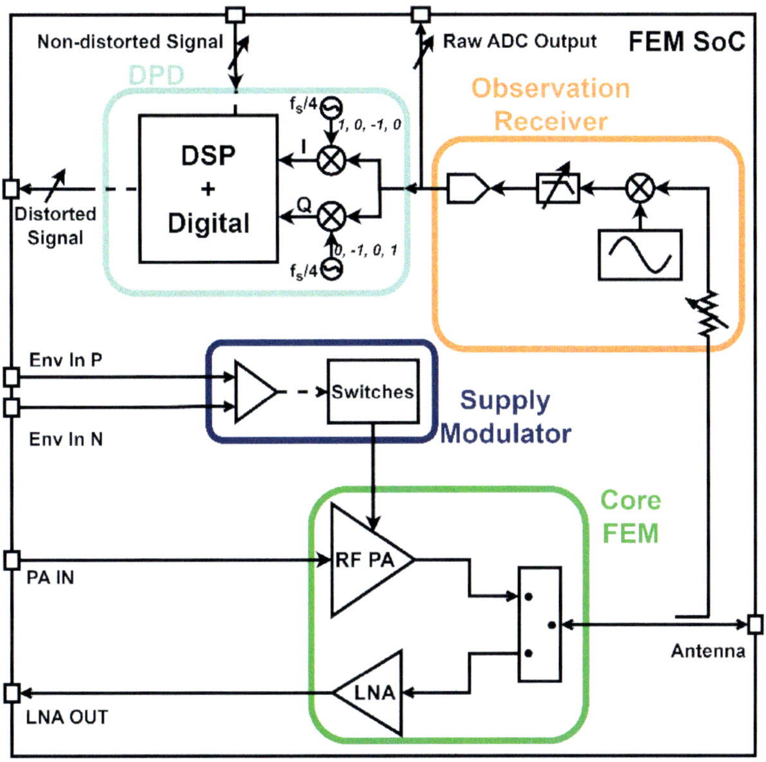

Fig. 1. System diagram of the V2X FEM SoC.

this reason, it is preferred to perform digital operations with the Fs/4 method.

Although the purpose of the feedback path is linearization, it will also have a manipulative contribution on the linearity of the sampled signal. A calibration method has been considered to eliminate this effect. After the calibration signal is applied to the observation receiver, the linearity of the feedback loop will be determined. One of the possible ways is to feed the calibration signal from the input of the observation receiver without using an extra input.

III. SUBSYSTEMS

Observation Receiver

The operating frequency band of the example system is 5855–5925 MHz. It has 10/20/30/40 MHz adjustable TX bandwidth supported by V2X standardization. Since, in order to perform DPD, it is necessary to receive signals with bandwidth of at least twice of the transmitted signal bandwidth, the observation receiver supports bandwidths up to 120 MHz. While there is a direct correlation between the receiver's bandwidth and DPD performance, ultra-wide bandwidths will result in wasted power consumption. For this reason, the receiver's bandwidth should be specified at a point suitable for the trade-off.

The "low-IF" topology is chosen as the receiver architecture. The most important factor taken into consideration when choosing the receiver architecture is to facilitate I/Q generation in the digital domain, because if the I/Q generation is realized in the analog domain, two ADC circuits will be needed. However, although this is not critical in terms of the area requirement, it can cause difficulties in terms of power consumption and number of package pins.

In Figure 2, the functional diagram of the input stage of the observation receiver is given. The observation receiver necessitates the use of an attenuator to address the wide range of input power levels. The purpose of the attenuator is to ensure that the power reaching the mixer remains at a consistently low level. By doing so, the attenuator prevents any negative impact on performance caused by the mixer's nonlinearities. The design incorporates 4-bit binary weighted resistive attenuators that offer both single and differential signal path options. To enable single to differential conversion, an on-chip balun is designed by using an EM tool. The attenuator allows for attenuation ranging from 30 dB to 42 dB, adjustable in increments of 0.75 dB through a digital input control word. While the design was optimized to achieve the desired attenuation specifically at 5.9 GHz, there may be slight variations in the attenuation as the frequency changes in band.

The down-converting mixer has a major effect on the linearity of the observation receiver. To enhance the dynamic range, it is essential to maintain low input power levels, which is why the preceding attenuator is employed. The double-balanced current-mode passive mixer is implemented. The source degenerated resistor in series with the mixer is connected to meet the linearity performance. The current-mode passive mixer exhibits the benefits of a significant conversion gain and linearity when compared to the conventional CMOS voltage mode passive down-conversion mixer. The mixer design shows a single-sideband integrated noise figure of 30 dB with a total conversion gain of 3 dB. The P1dB is obtained as 12 dBm in the mixer design.

The baseband filter design introduces an active-RC low-pass filter with linear-in-dB tunable gain and constant in-band group delay, incorporating a fast automatic cutoff frequency calibration. The baseband design comprises a fifth-order active-RC lowpass filter that incorporates a 4-bit programmable gain. The final stage of the filter incorporates a DCOC circuit, which effectively cancels out the output offset voltage. The automatic frequency tuning (AFT) circuit effectively adjusts the cutoff frequency to 120 MHz with default config, achieving an error rate of less than 4.7%. According to simulations, the frequency response and gain values of the baseband filter at various settings are given in Figure 3.

The pipeline ADC topology consisting of a sample and hold amplifier, 8 pipeline stages, a 2-bit flash sub-ADC, reference buffers, a clock generator, and digital error correction and retiming circuits is given in Figure 4. The input signal is first sampled by the sample and hold amplifier, which not only performs the sampling operation but also helps adjust the

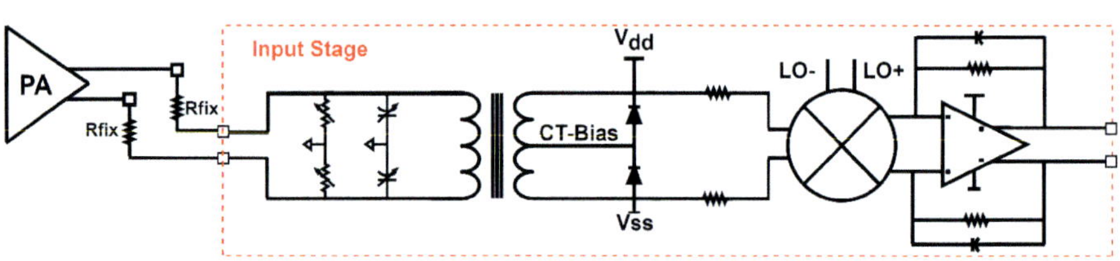

Fig. 2. Illustration of the input stage of the observation receiver.

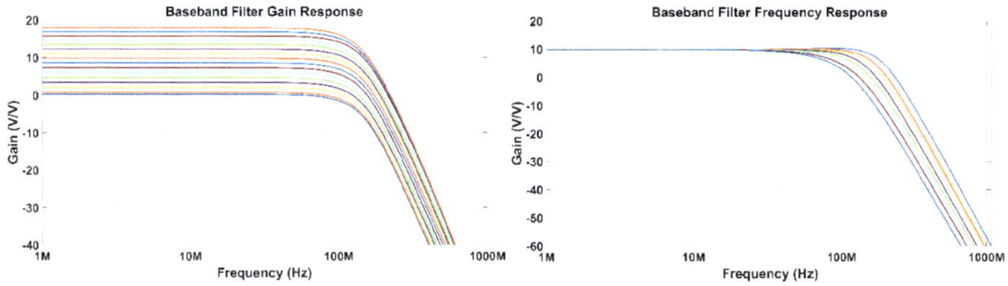

Fig. 3. Left: filter gain response for 120 MHz f_c, right: filter frequency response for 10 dB gain.

common mode of the output. The output of the sample and hold amplifier is then sampled by the first pipeline stage. Each pipeline stage consists of multiplying the digital-to-analog converter and sub-ADC which is a 1.5-bit flash ADC. Each stage performs the initial conversion with sub-ADC, providing a digital representation accurate to 1.5 bits that are stored in interstage registers. To minimize the impact of comparator noise, a 1.5-bit flash ADC is used instead of a 1-bit ADC. In the hold phase, the reference voltage produced by the reference buffer is subtracted from the sampled value and multiplied by two via the MDAC to produce a residual voltage. The last stage of the pipeline ADC is a 2-bit flash ADC. It takes residue voltage from last pipeline stage and convert it to 2-bit digital code. The digital outputs from each stage are combined and processed further to generate the 10-bit final digital output code via error correction and retiming circuit.

In the example system, a 10-bit, 250 MSPS pipeline ADC is designed and implemented. Spectrum analysis is done with simulations in order to measure the dynamic performance of the ADC and can be seen in Figure 5. The SNDR obtained from the simulation result is 62.25 dB at Nyquist frequency and the layout occupies 1.52 mm^2.

Digital Pre-distortion

In a DPD subsystem, an adaptive digital pre-distortion system with a closed-loop adaptation feature is implemented [1]. The developed DPD system is configurable to operate in the various modes with different options. These configurations are set from the SPI controller interface. The output of the analog observation receiver must be carried to the baseband to calculate the error signal for the adaptation, as shown in Figure 6. In order to remove the amplitude and the phase distortions created by the analog observation receiver, an equalizer circuit is used before the digital mixing stage. Constant amplitude distortions within

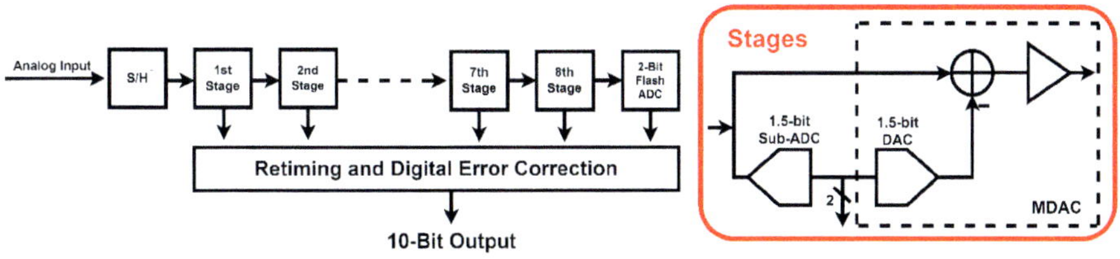

Fig. 4. Pipeline ADC.

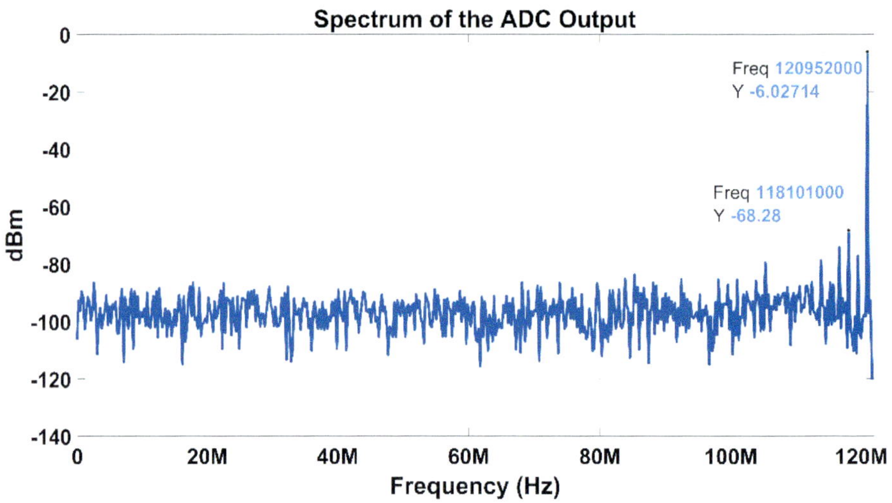

Fig. 5. Spectrum of the ADC output.

the analysis bandwidth can be corrected by the LUT itself since its content will be multiplied with the input signal of *u(n)*. Constant phase distortions within the analysis bandwidth are also corrected by using a phase rotation block, as shown in Figure 6. However, when these distortions are not constant within the analysis bandwidth, an equalizer block is required to compensate them. The equalizer circuit is designed as a 32-tap direct form FIR filter and its coefficients must be tuned for the observation receiver implementation.

After the equalizer stage, the IF signal is carried to the baseband by applying FS/4 digital mixing method as shown in Figure 1. Digital mixing is accomplished with the multiplication of [0, –1, 0, 1, …] and [1, 0, –1, 0, …] sequences with the digitized IF signal at the output of equalizer block to generate the in-phase and the quadrature signals. Also, the multiplication operation may be implemented using digital multiplexers and complement operators to reduce the implementation complexities further [2]. After the digital mixing stage, two 14-tap DDC FIR filters are designed to suppress the unwanted mixed copies of the IF signal.

At the output of DDC FIR filters, a complex baseband signal is generated and it is used to implement the digital pre-distortion algorithm. For the implementation, the LUT-based method proposed in [3] is used because of its fixed-point implementation in a single FPGA/ASIC device. The content of LUTs is adapted by using the error signal of *E(n)* and the input signal *u(n)*.

The latency between the baseband signal and the received signal is important for the power consumption and the complexity of the implementation. The total latency may become quite high when complex ADC and DAC serial interfaces with the following large digital filters are used at the transmitter and the receiver.

Supply Modulator

The supply modulator adaptively modulates the supply of the RF power amplifier for a TX signal of up to 40 MHz bandwidth. The supply modulator contains a fast AB class amplifier and switching amplifiers. The signal input to the supply modulator circuit is produced in analog

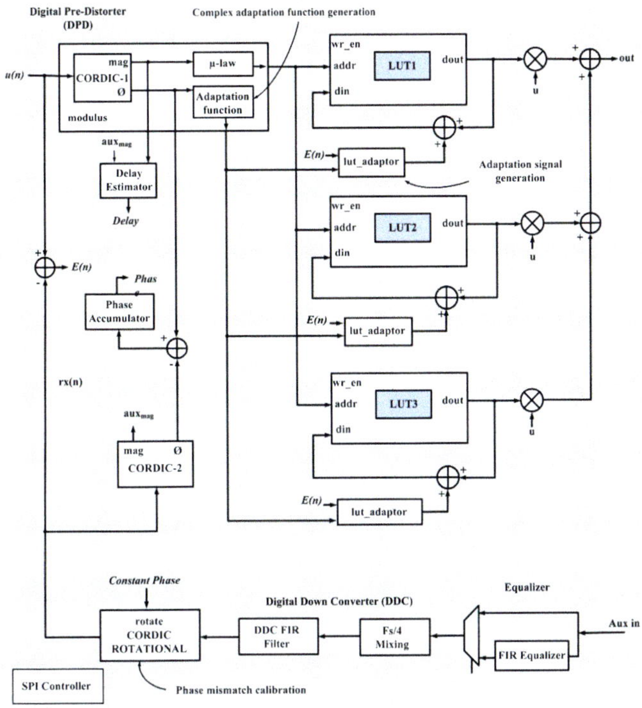

Fig. 6. Digital pre-distortion sub-system.

domain by using the baseband unit and external DACs. This signal is actually the envelope of the signal at the PA input.

Figure 7 demonstrates the AC-coupled variant of the parallel hybrid envelope tracking supply modulator. The output voltage of the linear amplifier is shifted by a large off-chip coupling capacitor. Consequently, using a reduced supply voltage for the linear amplifier results in an enhancement of the system's efficiency.

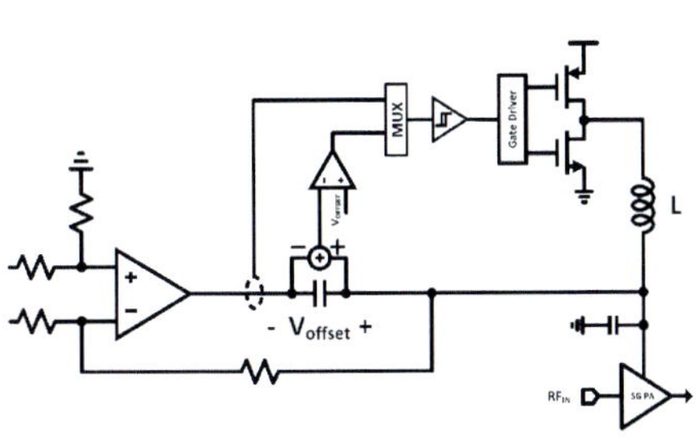

Fig. 7. AC-coupled hybrid supply modulator.

The supply modulator employing an AC-coupled technique has been designed to accommodate 64 QAM 5G NR signals with a maximum bandwidth of 40 MHz. The sub system has been simulated and Figure 8 illustrates the efficiency variations of the supply modulator across different levels of output power. The highest efficiency of 82.85% is achieved while the lowest efficiency of 75.38% is observed. As the output power increases, the impact of power consumption of the internal circuit on the efficiency of the supply modulator decreases.

Figure 8 also depicts the simulation result of output waveform of the AC-coupled envelope tracking supply modulator utilized for a 40 MHz, 64 QAM, 5G NR signal. In order to simulate the behavior of the RF power amplifier, a parallel circuit of a 4 Ω resistor and a 100 pF capacitor is employed. The graph depicts that the supply modulator exhibits the ability

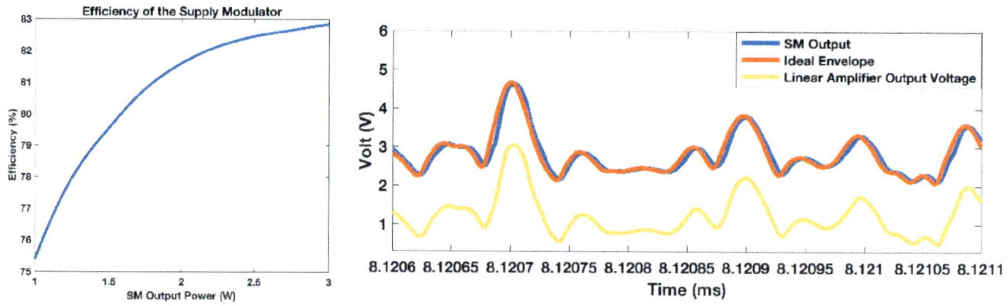

Fig. 8. Supply modulator (left) efficiency and (right) waveforms.

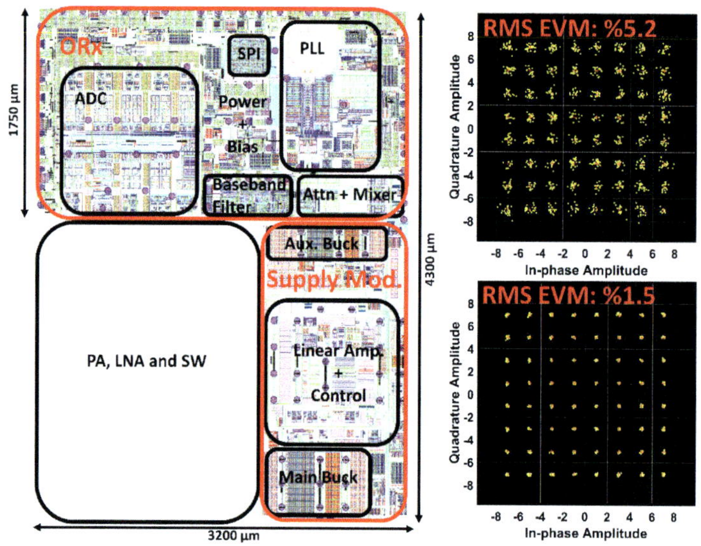

Fig. 9. Left: layout of the example SoC, right: EVM simulation results.

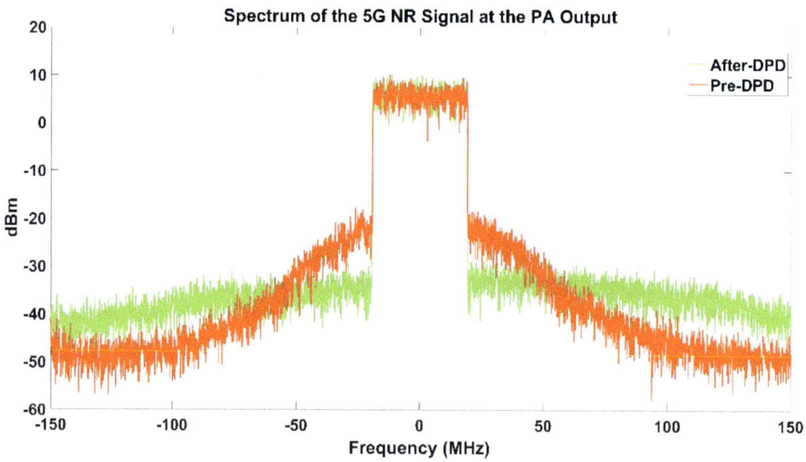

Fig. 10. The spectrum of the signal at the PA output.

to track the envelope signal without any limitations on the bandwidth.

IV. CONCLUSION

Assistance circuits that can be found in highly integrated RF front-ends for next generation communication systems are investigated. The example system is designed using the CMOS PD-SOI process and Figure 9 depicts its layout view. SoC has a size of 3200 μm by 4300 μm. At the time of this publication, the IC has been sent to production and is expected to be received from the foundry for testing.

The ORx has a noise figure of 77 dB at default settings including the attenuator. The RF PA has a model with memory, 28 dB gain, 33 dBm maximum output power with 4 V fixed supply and can be used with envelope tracking systems in a 2–5 V supply voltage range. After creating a realistic model of the ORx and DPD subsystem from post-layout simulations, the results of the EVM (error vector magnitude) simulation can also be seen in Figure 9. It can be deducted from the EVM results, the DPD circuit reduces the RMS EVM from 5.2% to 1.5%. The spectrum of the signal at the PA output with and without DPD is given in Figure 10. The Adjacent channel leakage ratio (ACLR) decreases from –30 dBc to better than –40 dBc for a 5G NR signal with 40 MHz bandwidth. Finally, the supply modulator, which can provide 2 A current up to 5 V output voltage, has over 82.8% efficiency for 40 MHz 5G NR signals with 9.8 dB PAPR.

REFERENCES

[1] D. Baran and E. Ulusoy, "A configurable implementation of adaptive digital predistortion system for RF power amplifiers," *2022 24th International Microwave and Radar Conference (MIKON), Gdansk, Poland*, 2022, pp. 1-4, doi: 10.23919/MIKON54314.2022.9924737.

[2] C. D. Presti, D. F. Kimball and P. M. Asbeck, "Closed-loop digital predistortion system with fast real-time adaptation applied to a handset WCDMA PA module," *IEEE Transactions on Microwave Theory and Techniques*, vol. 60, no. 3, pp. 604-618, March 2012.

[3] Y. Ma, Y. Yamao, Y. Akaiwa and C. Yu, "FPGA implementation of adaptive digital predistorter with fast convergence rate and low complexity for multi-channel transmitters," *IEEE Transactions on Microwave Theory and Techniques*, vol. 61, no. 11, pp. 3961-3973, Nov. 2013.

Emre Ulusoy received his B.Sc. and M.Sc degrees in electronics and communications engineering from Yildiz Technical University, Istanbul, Turkey in 2015 and 2018, respectively. He is doing his Ph.D. at Istanbul Technical University, Istanbul Turkey.

Since 2017, he has been working as a researcher at TUBITAK – BILGEM, Kocaeli, Turkey, leading the RFIC team. His current research includes integrated RF transceivers, mixed signal circuits, frequency synthesizers, data converters and high-speed wired and wireless communication circuits.

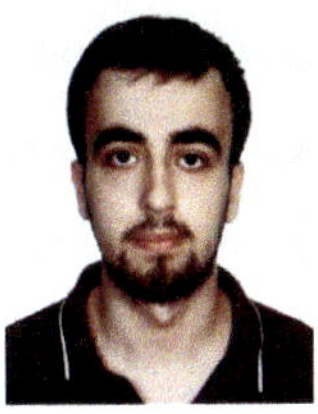

Bahadır Özkan received his B.Sc. degree in electronics and communications engineering from Yildiz Technical University, Istanbul, Turkey in 2019 and his M.Sc. degree in electronics engineering from Istanbul Technical University, Istanbul, Turkey in 2023. He works as an RFIC designer in TUBITAK – BILGEM, Kocaeli, Turkey. His research interests include analog and RFIC design.

Fatih Maden received his B.Sc. degree in electrical and electronics engineering from Middle East Technical University, Ankara, Turkey in 2019 and his M.Sc. degree in electronics engineering from the İstanbul Technical University, İstanbul, Turkey in 2023. Since 2019, he has worked at TUBITAK – BILGEM where he has been engaged in the research and development of mixed signal integrated circuits.

Furkan Barin received his B.Sc. and M.Sc. degrees in electronics and communication engineering from Yıldız Technical University and Istanbul Technical University, Istanbul, Turkey in 2020 and 2023, respectively. He is currently a researcher at TÜBİTAK – BİLGEM. His research interests include the design of analog and mixed-signal integrated circuits.

Ercem Yeşil received his B.Sc. and double major degrees from Istanbul Technical University, Electronics and Communications Engineering and Management Engineering Department, Istanbul, Turkey in 2021 and 2022, respectively. In 2022, he received a Fulbright scholarship award and is presently pursuing his Master's program in Electrical and Computer Engineering at the University of California, Los Angeles (UCLA). He was an analog/RF IC design engineer in TÜBİTAK – BİLGEM from 2021 to 2022. Currently, he is working as a mixed-signal design intern at Qualcomm. His primary research interests include the design of high-speed mixed-signal circuits in CMOS technologies.

Adem Eren received his B.Sc. degree in electronics and communication engineering from Istanbul Technical University, Istanbul, Turkey in 2020. He started his M.Sc. study at Gebze Technical University, Kocaeli, Turkey in 2022. Since 2020, he has worked at TUBITAK – BILGEM as a researcher. His research interests include processor architecture, artificial neural networks, and communication protocols.

Dursun Baran (S'09) was born in Tokat, Turkey, in 1984. He received his B.Sc. degree in electrical engineering from Bogazici University, Istanbul, Turkey in 2007, and his M.S.E.E. degree from The University of Texas at Dallas, Richardson, TX in 2009. In 2007, he joined the Advanced Computer Systems Engineering Laboratory (ACSEL), The University of Texas at Dallas, where he worked in the area of high performance and low power digital circuits. His research interest focus was on architectural optimizations for energy efficient and high performance arithmetic circuit blocks. He completed his doctoral study at the same university in 2011. He proposed various optimization techniques to increase the energy efficiency of high performance digital circuits in his doctoral work. In 2011, he joined the Informatics and Information Security Research Center (BILGEM) of Turkey to conduct research on high speed digital signal processing algorithms and circuits. He implemented various DSP algorithms in FPGA environment for defense and aerospace applications. He then moved to IC Training and Design Laboratory (TUTEL) in BILGEM to work on mixed signal IC design and optimization research.

Tufan C. Karalar received his BS. Degree from University of Michigan, Ann Arbor in 2000, MS and PhD degrees from University of California, Berkeley in 2002 and 2006 respectively. From 2006 to 2010 he was a staff design engineer working at Broadcom Corp. working on data converters, analog mixed signal circuits, filters, sensor interfaces. From 2010 to 2014 he was a Sr. Design engineer at Silicon Laboratories Inc. working on RF circuits, Analog mixed signal circuits, power electronics and digital circuit design. Since 2015 he has been with Istanbul Technical University Electronics and Communications Engineering Department first as an Assistant then as an Associate Professor. His research interests include, Sensor interface ASICs, High Speed Data converters and digital circuits, Power management IC's, Sensors in emerging technologies. He has 4 issued and 4 pending patents.

Ahmet Tekin has received his EE Ph.D. degree from University of California Santa Cruz, CA, his EE M.Sc. degree from North Carolina A&T State University, Greensboro, NC and his EE B.Sc. degree from Bogazici University, Istanbul, Turkey in 2008, 2004, and 2002, respectively. In addition to academic research in microelectronics, he has worked for multiple innovative semiconductor design companies such as Multigig, Inc., Newport Media, Aydeekay LLC, Broadcom corp., Semtech Corp., Nuvoton Technology Corp., Qualcomm, Pharrowtech and Waveworks Inc., leading semiconductor designs for communications, consumer and medical markets. He has also been active in academia and served as a faculty member in Özyeğin University, Istanbul Technical University and most recently Boğaziçi University. His main focus area is analog/RF/mixed-signal integrated circuit design for sensing, communication and biomedical applications. He is currently engineering faculty member of Boğaziçi University Department of Electrical and Electronics Engineering.

Ertan Zencir received his B.Sc. and M.Sc. degrees in Electrical and Electronics Engineering from Middle East Technical University in Ankara, Turkey in 1995 and 1997, respectively. He completed his Ph.D. in Electrical Engineering from Syracuse University in Syracuse, NY, USA, in 2003. After completing his doctoral studies, he worked as an assistant professor at the University of Wisconsin-Milwaukee in Milwaukee, WI, USA from 2004 to 2005. He then held the position of senior RFIC design engineer at Nokia Mobile Phones RFIC group in San Diego, CA, USA from 2005 to 2006, focusing on CMOS/BiCMOS cellular RF transceivers. From 2006 to 2011, he served as a principal design scientist in the broadband RF group at Broadcom Inc in Irvine, CA, USA specializing in RF and analog designs for satellite receivers. Since 2012, he has been an assistant professor in the Electrical and Electronics Engineering Department at the University of Turkish Aeronautical Association in Ankara, Turkey. Additionally, since 2017, he has been the technical lead of the RFIC team at TÜBİTAK – BİLGEM Research Center in Gebze, Turkey. His research interests primarily revolve around the design of wireline and wireless RF and analog integrated circuits for advanced communication systems.

Impact of a High-resistivity Substrate on RF and mm-wave Performance of 22 nm FD-SOI Devices and Circuits

Martin Rack, Lucas Nyssens, Massinissa Nabet, Dimitri Lederer, and Jean-Pierre Raskin

***Abstract*—This paper studies the impact of silicon-substrate bulk and interface resistivity properties on the performance of mm-wave devices and circuit modules implemented in 22 nm FD-SOI. To achieve this, key devices were fabricated in GlobalFoundries's 22FDX® node and studied on several types of substrates, from the conventional 10 Ωcm silicon wafers with over 500 Ωcm high-resistivity ones. Furthermore, several silicon-interface conditions were explored, offering process variations in the substrate's interface resistivity on top of the change in substrate bulk resistivity. First, substrate influence on passive structures are investigated using CPW lines and spiral inductors, and significant improvements in lineic loss and quality factor are demonstrated when using improved high-resistivity (>500 Ωcm) substrate materials and passivated interfaces over the conventional reference process using standard resistivity silicon (10 Ωcm). Next, several key front-end circuits were designed, fabricated and measured to demonstrate the substrate impact on higher complexity mm-wave modules, including switches, low-noise amplifiers (LNA) and power amplifiers (PA). The insertion-loss of the designed DC-40 GHz switch is measured to improve by 0.35 dB when employing the high-resistivity substrate options. The LNA demonstrator circuit operates from 42 to 67 GHz with a gain and noise figure of 19.90 and 3.76 dB at 55 GHz, which are shown to improve by +0.95 dB and –0.25 dB respectively when moving to the low-loss substrate option. Finally, the PA demonstrator circuit, which operates from 24 to 43 GHz with a peak power added efficiency of 25.1% and output power of 17.8 dBm on the reference standard substrate has those metrics improved to 26.6% and 18.1 dBm on the high-resistivity wafer. This work gives an overview of the increase in performance levels of the main key mm-wave passive and circuit blocks, and demonstrates that the substrate impacts the main figures of merits of these devices by 0.3 to 1.0 dB in the 20–60 GHz range of the devices considered in this study.**

***Index Terms*—RF CMOS, silicon-on-insulator (SOI) technology, fully depleted (FD) SOI, RF and millimeter-wave performance, silicon-based substrate, high-resistivity substrate, coplanar waveguides (CPW), inductors, RF switches, LNA, PA.**

The authors would like to thank GlobalFoundries for chip fabrication and research support. This work is supported by Ecsel JU project "BEYOND5" (grant agreement No 876124) via EU H2020 and Innoviris (Brussels/Belgium) funding.

Martin Rack, Lucas Nyssend, Massinissa Nabet, Dimitri Lederer and Jean-Pierre. Raskin are with Université catholique de Louvain (UCLouvain), 1348 Louvain-la-Neuve, Belgium (e-mail: martin.rack@uclouvain.be).

I. INTRODUCTION

NOWADAYS, advanced CMOS nodes are competitive for radio-frequency (RF) and mm-wave applications. In particular, fully-depleted

(FD) -SOI devices boast f_t and f_{max} metrics in the range 300–400 GHz [1], and support rich back-end-of-line (BEOL) options with up to 11 metal layers, including several thick copper layers supporting high-Q RF/mm-wave passives and interconnects.

The silicon substrate is part of the electromagnetic environment of such passives and devices, and can be responsible for significant amounts of losses, coupling and non-linear signal distortion [2] if the silicon resistivity is low.

Using high-resistivity (HR) silicon substrates (with nominal resistivity ρ_{nom} > 500 Ωcm) rather than standard-doped silicon (ρ_{nom} = 10 to 20 Ωcm) brings an improvement to these effects, though the benefits are hindered by the *parasitic surface conduction* (PSC) effect. A PSC layer is induced by fixed positive charges at the Si/SiO_2 interface that attract significant amounts of free electrons, forming a thin channel-like layer that is highly conductive [3, 4]. PSC then degrades (lowers) the *effective resistivity* (ρ_{eff}) sensed by coplanar circuitry overlying the substrate stack. For these reasons, the ρ_{eff} of HR substrates are typically only in the range 30–150 Ωcm [2, 5].

Interface passivation solutions have then been developed to counter the PSC, the most widespread of which is the *trap-rich* solution, that introduces a thin layer of polysilicon between the SiO_2 and Si, that is rich in defects that trap free carriers and render the interface resistive [4, 5, 6]. Trap-rich SOI has seen strong industrial success in partially depleted (PD)-SOI nodes. However, FD-SOI supports below-BOX features (back-gate contacts, isolations wells, diodes, etc.) that are difficult to integrate within defect-rich poly-Si. Such compatibility issues motivate the development of *alternative interface passivation schemes*. For these reasons, this paper will study in particular the novel *PN-interface passivation* technique applied to FD-SOI.

In this paper, the impact of the substrate and its interface passivation's properties on the figures of merit of mm-wave devices are studied. This study was conducted using GlobalFoundries' 22 nm FD-SOI node, and the designed circuits were all fabricated and measured on different substrate options, ranging from standard-resistivity wafers to high-resistivity silicon, and exploring different interface passivation solutions.

First, the substrate's impact on key passive components such as coplanar waveguide (CPW) transmission lines and spiral inductors are presented. Then, the substrate's impact on key mm-wave front-end circuit blocks are described. In particular, studies are performed – supported by measurement data – of the substrate's influence on (i) the insertion loss of DC-40 GHz switches, (ii) on the saturated output power and the power-added efficiency (PAE) of a 24–43 GHz power amplifier (PA) design, and (iii) on the noise figure (NF) and gain of a 42–67 GHz low-noise amplifier (LNA).

In this work, GlobalFoundries' 22FDX® technology was run on several different types of FD-SOI wafers (p-type) with various bulk silicon resistivity values. Furthermore, three different interface variations are considered: (i) with the default, i.e. process of record (POR) interface, (ii) with an "enhanced" interface, and (iii) with a *PN interface passivation technique*. These different options are not part of GlobalFoundries' standard technological offerings, and are part of a unique and exploratory process.

Those overall six substrate options (two interface possibilities and four substrate bulk-resistivity options) are depicted in Figure 1, along with a representation of the technology's back-end of line (BEOL) stack. At the "enhanced interface", special

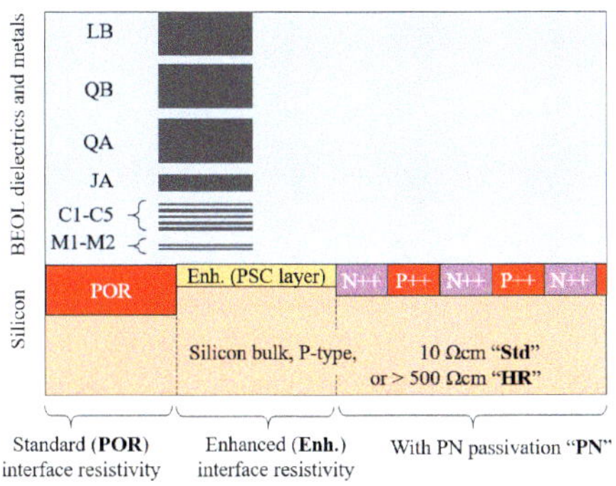

Fig. 1. Depiction of the different substrate options (in the bulk and at the interface) and the overlying BEOL metal stack.

care was taken to avoid any local impurities, and therefore any conductive region present in that case is associated with a PSC induced by fixed interface charges.

The concept of the PN-junction substrate was first proposed and explained in [7, 8] and implements alternating regions of P- and N-type doping to locally interrupt the would-be conductive interface (PSC effect) by induced depletion junctions, whose chain-series combination results in a strong increase in overall substrate impedance (ρ_{eff}) sensed by overlying coplanar circuits. These depletion zones interrupt the conductive sheet because they are inherently highly resistive, and thereby serve to increase the effective sheet resistivity at the Si/SiO_2 interface.

This chapter further adds to those studies on PN interface passivation as the first work to present the impact and benefits on full FEM circuit modules, including a wideband single-pole double throw (SPDT) switch, a 42–67 GHz LNA and a 24–43 GHz PA, all designed in 22 nm FD-SOI.

This chapter is organized as follows. First, in Section 2, the different substrate options used in the unique 22FDX® run of this work are described. Section 3 reports the impact of the substrate option on the loss of coplanar waveguide designs in implementing the topmost copper metal layer QB. The effective resistivity sensed by these lines is also extracted from the measured data and gives strong insight into the actual bulk resistivity and quality of interface passivation of the different wafers. In Section 4 the results of spiral inductors are presented, also implemented in the QB metal layer, and the substrate's impact on their quality factors are demonstrated. Section 5 discusses the substrate impact on the figures of merit of mm-wave DC-40 GHz single-pole-double-throw (SPDT) switches, with the main impact being on the observed insertion loss. It is demonstrated through both measurements and simulation data that the insertion-loss at 30 GHz can be improved from 2.10 dB to 1.75 dB by simply modifying the effective substrate properties below the SPDT. The substrate's effect on a 42–67 GHz LNA are then studied in Section 6, and an increase of 0.95 dB in gain and decrease of 0.25 dB in noise-figure is demonstrated at 55 GHz by using the high-resistivity substrate over the reference standard resistivity. Finally, Section 7 presents the substrate's impact on a 24–43 GHz power amplifier design, and an increase of from 25.1% to 26.6% is observed in the peak power added efficiency (PAE) of the amplifier, along with an 0.3 dB increase in the output power (at the maximum PAE point).

Overall, meaningful gains are observed in this variety of mm-wave integrated circuitry, and a recap of the substrate impact on them all are given with the conclusion in Section 8.

II. 22FDX® SUBSTRATE VARIATIONS

Figure 1 depicts the six substrate variations atop which the CPW lines and the mm-wave FEMs of this work were fabricated.

At the Si/SiO_2 interface, three different options were considered:

- *POR*: this is the standard process-of-record (POR) option in the 22FDX® line, and the interface resistivity is set during certain steps of that reference process. This interface option will be labelled as "POR" from here on.
- *Enhanced*: in this option, the aforementioned reference process is optimized to yield an interface that is less conductive. A fairly conductive PSC effect is still present in this case, but that remains less conductive than in the POR case. This interface option will be labelled as "Enh" from here on.
- *With PN passivation*: in this option, the enhanced process is applied (instead of the POR), and, on top of that, P^{++} and N^{++} doped regions are defined in alternating patterns beneath the RF devices. These regions are defined through implantation process steps using the same implant parameters (impurities, energies and doses) as those used for the back-gate well definitions of this FD-SOI technology. Furthermore, on-chip interconnects to each P and N region have been included to offer DC-biasing of the PN junctions. This interface option will be labelled as "PN" from here on if no DC bias is applied to the PN-junctions, and as "PN3V" if a reverse junction bias of 3 V is enforced.

In practice, each P and N region has DC bias applied to it through large-valued polysilicon resistors implemented on-chip. The point is that, by applying a reverse bias to the PN junctions the depletion regions can be widened, and, since these regions dominate the sheet impedance, this increases the effective sheet resistivity at the interface. All three of the above interface options have been implemented and fabricated on two types of base wafers in the 22FDX® process line:

- *Standard resistivity*: this bulk silicon has a resistivity in the range 10–20 Ωcm. This substrate option is currently part of the reference process in the 22FDX® line, and will be labelled as "Std" from here on.
- *High resistivity*: using this option, the bulk silicon has a resistivity in the range 500–1000 Ωcm. This substrate will be labelled as "HR" from here on.

III. SUBSTRATE IMPACT ON CPW LINES

Figure 1 depicts the metallic layers of the BEOL in the fabricated chip of the devices considered in this work. It comprises seven thin (thicknesses in the range of 100 nm) copper layers, M1, M2 and layers C1 to C5. On top of those comes a 1 μm-thick copper JA layer, followed by two 3 μm-thick copper QA and QB layers. Finally, the technology is capped-off with an aluminum 2.9 μm-thick LB layer.

A set of CPW lines was fabricated in the thick high-level metal QB to showcase the typical high performance that is achievable in terms of losses on the different substrates from CPW lines in the technology. To achieve 50 Ω of characteristic impedance (Z_c), the central signal line's width (W_c) was set to 35 μm, and the spacing (S) to the outer ground lines was set to 20 μm.

Lines of two different lengths were measured (760 and 2000 μm) along with dedicated Thru (290 μm of total length) and Open structures.

The fabricated CPW lines were measured on-chip up to 67 GHz using a pair of GSG probes with 100 μm pitch from FormFactor and a PNA-X analyzer from Keysight. Following the extraction procedure described in [9], the pad parasitics are de-embedded and an mTRL approach is applied to extract the properties of the uniform line, such as the real and imaginary parts of both the propagation constant and the characteristic impedance, ultimately leading to the evaluation of the equivalent lineic RLGC parameters. Following that extraction, the CPW lines on the various substrates are compared in terms of lineic losses (α) and substrate effective resistivity (ρ_{eff}). The latter is computed from the lineic G parameter thanks to a conformal mapping technique of CPW geometry [10], as detailed in [9].

The extracted data pertaining to the QB CPW lines is given in Figure 2. The plotted curves highlight the substrate impact on the overall performance of the CPW line, which exhibits a loss of 1.00 dB/mm at 30 GHz on the Std-POR substrate.

It is demonstrated that this loss can already be substantially reduced by including the enhancement process of the interface resistivity compared to the POR reference process, with the same CPW on the Std-Enh wafer exhibiting 0.75 dB/mm of loss at 30 GHz. This reduction is explained by the higher effective resistivity sensed by the line, which is increased from 35 Ωcm to 47 Ωcm due to the less conductive interface.

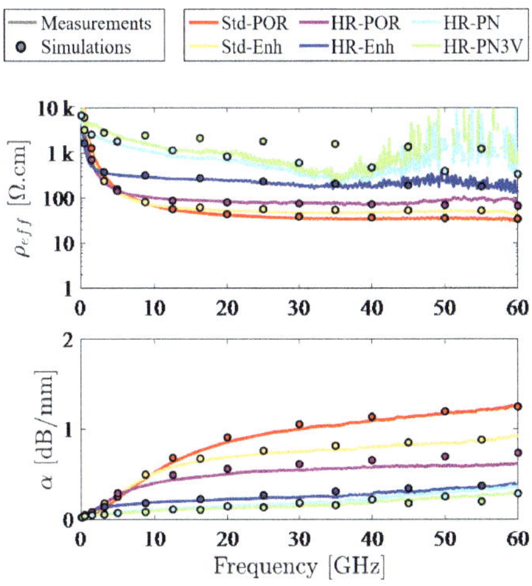

Fig. 2. Effective resistivity (ρ_{eff}) and line losses (α) from the measured CPW lines (QB metal layer) on the different considered substrate options.

On the HR substrate, the lines on the HR-POR sample exhibit losses of 0.54 dB/mm at 30 GHz. Compared to the Std-POR substrate, 0.46 dB/mm of loss have been reduced by simply modifying the base resistivity of the wafer, with the effective resistivity increasing from 35 Ωcm to 78 Ωcm at 30 GHz. This loss is further reduced by using the enhanced-interface process option, with the lines on the HR-Enh sample presenting lineic losses of 0.25 dB/mm for an effective resistivity value of 196 Ωcm at 30 GHz. In that case, the interface resistivity is determined by a PSC effect present at the Si/SiO_2 interface, which dominates the G term of the line parameters. This interface resistivity can be increased further by utilizing the PN junction passivation technique. This is demonstrated by the further reduction in the line-loss, down to 0.18 dB/mm, when implementing the CPW lines on the HR-PN sample.

A frequency-dependence is observed in the measured G and ρ_{eff} data, as expected, from the series combination of conductive doped regions (N or P) with capacitive depletion regions [7]. The depletion regions can be extended in space by reverse-biasing the PN junctions. This has the effect of increasing the interface impedance and reducing further the line loss, as is demonstrated by the CPW data labelled as HR-PN3V. These lines are identical to the HR-PN lines, with the addition of a 3 V biasing applied to the PN patterns.

Overall, lines on the HR-PN and HR-PN3V samples achieve ρ_{eff} values of 360 and 457 Ωcm at 30 GHz for line losses of

0.18 and 0.15 dB/mm respectively. Simulation results fit well to the measured data and reveal that 0.13 dB/mm of the line-losses at 30 GHz are due to the finite metal resistivity of the QB layer, meaning that the substrate's loss contribution on the HR-PN3V substrate is very low at around 0.02 dB/mm, confirming excellent overall substrate passivation in that case.

IV. SUBSTRATE IMPACT ON SPIRAL INDUCTORS

Spiral inductors of 1 nH also implemented atop the different substrate options are measured from 100 MHz to 67 GHz. The inductance (L) and quality factor (Q) of the inductors are computed as $L = \mathrm{Im}(1/Y_{11})/\omega$ and $Q = \mathrm{Im}(1/Y_{11})/\mathrm{Re}(1/Y_{11})$. Figure 3 shows the 1 nH inductance and quality factor of the inductor on several substrates.

The inductors chosen to demonstrate the substrate's impact in this work have 2.5 turns implemented in the QB layer, with an inner diameter of 72.5 μm, a coil width of 6 μm and an inter-coil spacing of 5 μm. These dimensions are highlighted in the inset of Figure 3.

These inductors were fabricated on the same aforementioned substrates, both with and without a resistive bulk option,

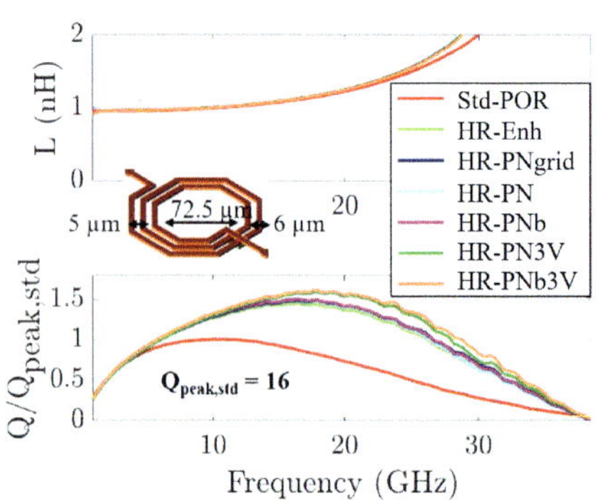

Fig. 3. Inductance (L) and normalized quality factor (Q) of 5–20 GHz inductors on different substrates, with or without interface passivation.

as well as with the option of several interface passivation solutions. The same PN patterns as below the CPW lines were implemented (labelled as "-PN" in Figure 3), but also other types of PN pattern shapes were investigated beneath these inductors, including an array using different geometries (annotated as "-PNb" in Figure 3) and a grid array (annotated as "-PNgrid"). Those results have been previously reported in [11], and the interested reader can find the details of the PN implants in that work. The quality factor in Figure 3 is normalized by the peak value of Q on the Std-POR substrate ($Q_{peak,std} = 16$).

The RF inductor quality factor is greatly improved as we move from that standard substrate to a HR-Enh substrate thanks to the greater bulk resistivity. Indeed, the peak Q value is around 1.5 times higher (from 16 to 24) and shifted to higher frequencies. Further improvement (from 24 to 26) is achieved thanks to the 3 V reverse biased PN junctions, which significantly increases the substrate resistivity at the interface at high frequency.

V. SUBSTRATE IMPACT ON A MM-WAVE DC-40 GHZ SPDT

SPDT Module Design

A DC-40 GHz SPDT module was designed and fabricated on the same Std and HR wafers as the CPW lines in the 22FDX® process from GlobalFoundries.

A classic series-shunt topology is chosen for wideband performance, the layout and schematic of which are illustrated in Figure 4. The design utilizes 20 nm-gate-length super-low threshold voltage (SLVT) devices that include back-gate contacts to offer the lowest $R_{on}C_{off}$ metric in the technology. In each branch of the switch, three FETs are stacked in series to achieve a P1dB close to 20 dBm. After that, the FETs are sized to achieve the lowest possible insertion loss (IL) while

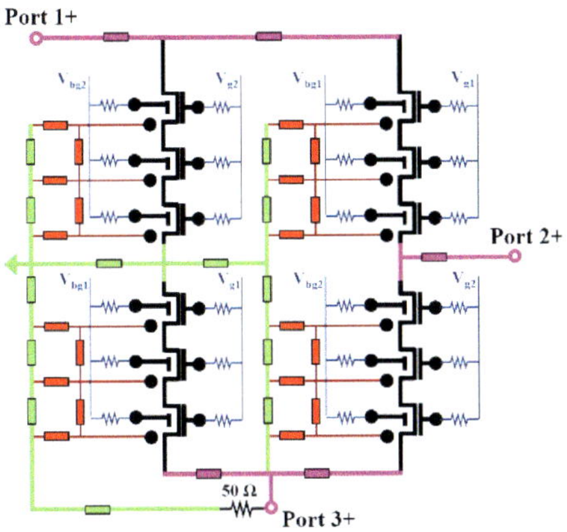

Fig. 4. Mm-wave SPDT switch design.

maintaining better than 20 dB isolation (ISO) from DC to 40 GHz.

Through this design methodology, the transistor widths in the series branches were determined as 150 μm, and those in the shunt branches were determined as 33 μm.

All FETs are multi-finger devices using a finger width of 3 μm, and have their front- and back-gates biased through on-chip 10 kΩ resistors. The four DC control voltages of the switch, depicted in Figure 4, are V_{g1}, V_{bg1}, V_{g2} and V_{bg2}. In the on-state, which connects Port 1 to Port 2, those four voltages are set to 0.9 V, 3 V, –0.9 V and 0 V, respectively. In the off-state, which connects Port 1 to Port 3 (and isolates port 1 from port 2), those four voltages are set to –0.9 V, 0 V, 0.9 V and 3 V, respectively. Finally, Port 3 is implemented on-wafer using an integrated precision 50 Ω load. A microphotograph of the fabricated DC-40 GHz SPDT module is given in Figure 5.

Measurements of Substrate Impact on Switch FoMs

The SPDT switch implemented on each of the substrates is both measured and simulated fully with the reference planes set on the GSG probe pads. An open-pad

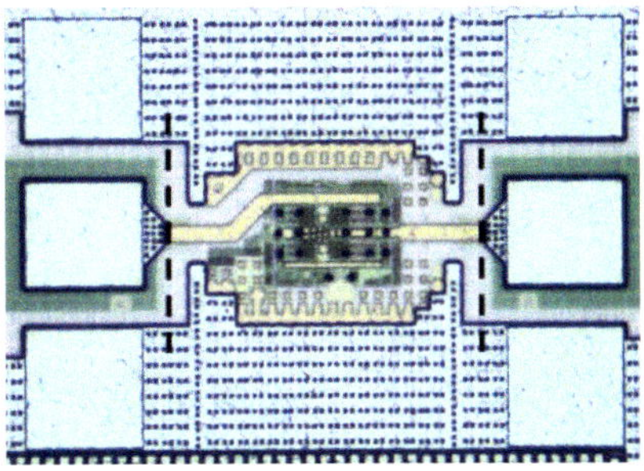

Fig. 5. Microphotograph of the fabricated SPDT module. The measurement reference planes after open-de-embedding are shown as dashed lines.

de-embedding procedure is applied to both the measurements and simulations of the full SPDTs, by which the Y-matrix of a measured (simulated) open-pad structure is subtracted from the Y-matrix of the measured (simulated) full SPDT.

The resulting data is then plotted in Figure 6, that gives the S-parameter results after open-de-embedding in both the on- and off-states of the switch up to 60 GHz.

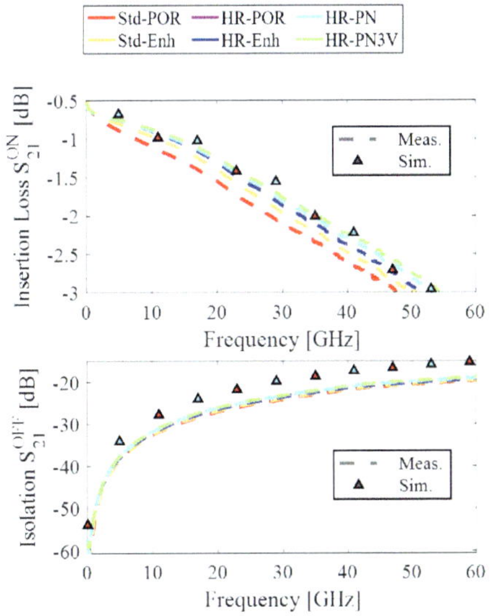

Fig. 6. Measured and simulated S_{21} parameter of the SPDT (open-de-embedded) as a function of frequency on all substrates in both the on- and off-states.

The switch meets the design specifications by offering better than 20 dB isolation (S_{21}^{OFF}) up to 40 GHz. In the on-state, the switch on the reference Std-POR substrate exhibits an insertion loss of 2.10 dB at 30 GHz. This can be improved to 1.97 dB by enhancing the interface resistivity (Std-Enh). By simultaneously enhancing the interface properties and increasing the bulk resistivity, the insertion loss is improved to 1.86 dB for the HR-Enh sample. Adding the PN-junction passivation scheme on top of that further decreases the 30 GHz loss to 1.80 dB (HR-PN), which can be reduced to 1.75 dB by applying 3 V of reverse junction-bias (HR-PN3V).

Figure 7 highlights those gains achieved by increasing the substrate impedance on the insertion loss. Overall, 0.35 dB improvement is achieved in the switch's insertion loss at 30 GHz, offering loss values as low as 0.87, 1.24 and 1.75 dB at 10, 20 and 30 GHz, respectively. This is achieved in practice thanks to the increased impedance brought to the substrate network parasitics (shown to deteriorate switch performance in FD-SOI technology through simulation analysis in [12, 13]) by the combination of the high-resistivity bulk and the optimizations of the interface properties.

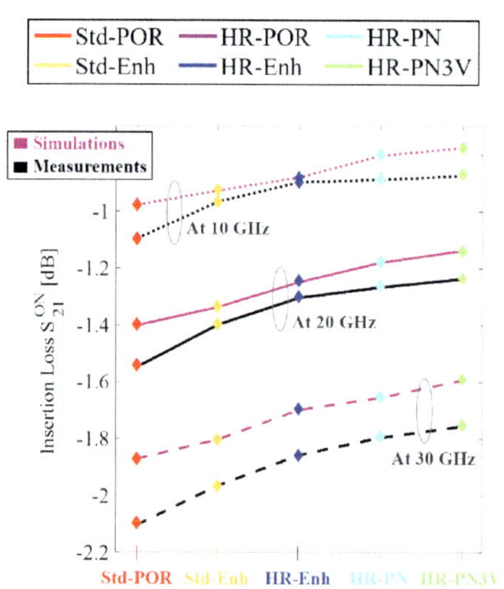

Fig. 7. Measured and simulated SPDT insertion loss (open-de-embedded) on the different substrate options.

VI. SUBSTRATE IMPACT ON A MM-WAVE 42-67 GHZ LNA

LNA Design

The overall schematic along with the core layout of the three-stage LNA is presented in Figure 8. The LNA design includes three stages, with each stage being a cascode topology with degenerated source inductor (L_s) with a transformer load (L_g and L_d). The multistage cascode architecture is chosen to realise high gain while maintaining a low power consumption in this technology, and the transformer-loaded technique is used to boost the gain and trade-off gain for bandwidth [14].

All transistors in the LNA are SLVT N-FETs of 20 nm gate length, and are sized with 500 nm finger width, which provides a good trade-off point between available minimal noise figure (NFmin) and available gain. Each transistor is made up of four transistor cells of 5 μm width each, using 10 fingers, using a bias condition of 0.8 V of source to drain voltage (V_{ds}) on each FET (therefore a V_{dd} of 1.6 V) and a current of 4.2 mA per stage (corresponding to a current density in each device of 0.21 mA/μm, which is at the minimum noise figure performance).

The degeneration inductors L_s at the common-source FETs are included to help stabilize the amplification stages and to tune the input impedance of the first state such that matching can be achieved using a single series input spiral inductor L_{in}. This approach minimizes the complexity of the input matching network prior to the first amplification stage, thereby minimizing the impact of that network on the overall noise figure of the LNA.

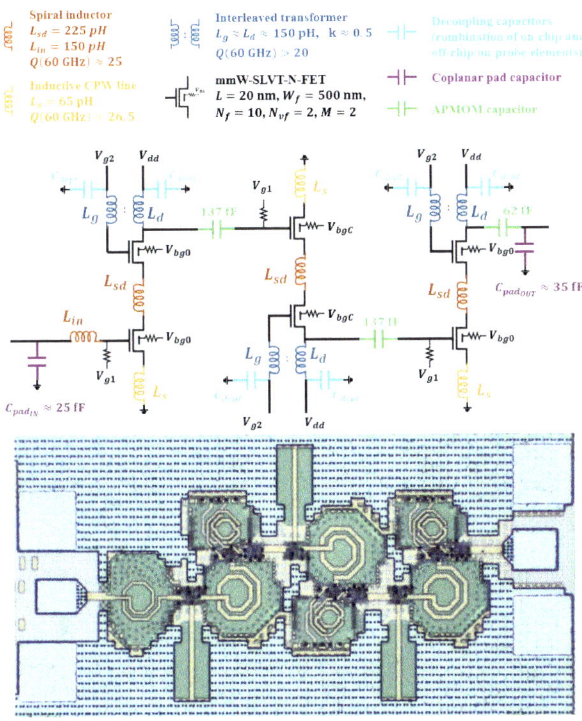

Fig. 8. Schematic of the three-stage LNA design and microphotograph of the fabricated circuit.

Inductors L_{sd} are placed in between the common-source (CS) and common-gate (CG) transistors in each amplification stage in order to resonate-out capacitive parasitics at the node between these FETs and to suppress the noise contribution of the CG transistor.

Inductors L_d are included at the drain nodes of the CG devices to resonate-out the drain capacitances and peak the gain, while inductors L_g are included at the gate node to increase the cascode gain of each stage. However, large values of L_g tend to render the CG stage unstable, but this is compensated for by introducing coupling between L_g and L_d [15]. During the design, L_g is introduced to boost the gain, and then coupling is introduced with L_d to achieve stability. Furthermore, L_g and k can be tuned to trade-off peak gain at a single frequency versus a flatter wider-band response of the gain [15]. This was exploited in this work, where it was found that using values of L_g and L_d of 150 pH along with a coupling factor of 0.5 between them that a wideband response could be achieved for the amplifier, from 42 to 67 GHz.

Measurements of Substrate Impact on LNA Performance

The fabricated LNA was measured on-wafer using two types of setup. For the noise-figure characterization, a 2 Hz to 110 GHz N9041B UXA signal analyser from Keysight was employed along with a 60 GHz noise source. For the gain measurement, S-parameter measurements were performed using a 500 Hz to 125 GHz N5291A vector network analyser from Keysight.

The LNA was biased using two six-pin DC probes placed in the north and south configuration at 90° to the RF probes through dedicated DC pads. These probes include decoupling capacitors that, along-side on-chip integrated decoupling elements, serve to enforce a wideband decoupling between the V_{dd} pins and the GND pins of the LNA, and in this configuration all stages of the LNA are unconditionally stable over a wideband (verified from DC to 110 GHz).

The S-parameter and noise-figure measurement results are plotted Figure 9 under the nominal bias condition. At this bias the three-stage LNA draws a DC power of 12.5 mW. The measurement results demonstrate a wideband operation of the LNA from 42 to 67 GHz. On the Std-POR substrate, a peak gain of 20.5 dB is achieved along with a minimal noise-figure value of 3.7 dB.

Simulation results predict that, by moving the LNA design to a higher-resistivity substrate, a 0.25 dB decrease in the noise figure and a 0.95 dB increase in gain can be achieved. The noise figure improvement is mainly attributed to reduced losses in the input matching network (primarily in inductor L_{in}), and the increase in gain is attributed to reduced losses throughout the passive devices of the entire LNA (mainly all spiral inductor elements, that see an appreciable improvement in quality factor, as discussed and highlighted in section 4).

VII. SUBSTRATE IMPACT ON A MM-WAVE 24–43 GHZ PA

As a PA demonstrator design, a balanced differential multi-stage topology based on a distributed output balun [16] was chosen to achieve wideband performance over the n257-260 bands with high output power. For VSWR aspects, a balanced topology offers an inherent robustness to load variations, with increased output power [17].

The designed PA relies on a three-stacked FET output stage with a distributed balun, covering 24–43 GHz. The PA is based on a pseudo-differential two-stage configuration, with the schematic shown in Figure 10. The power (second) stage uses a three-stacked FET differential architecture [18] while the driver (first) stage features no FET stacking. Their circuit schematics are shown in Figures 10b and 10c, respectively. Both stages use cross-coupled neutralization capacitors to stabilize the amplifier in differential operation [19]. The single-ended to differential (and inversely) transformation is performed by the input and output baluns that also implement the required impedance matchings. An inter-stage transformer transforms the input impedance of the power stage into

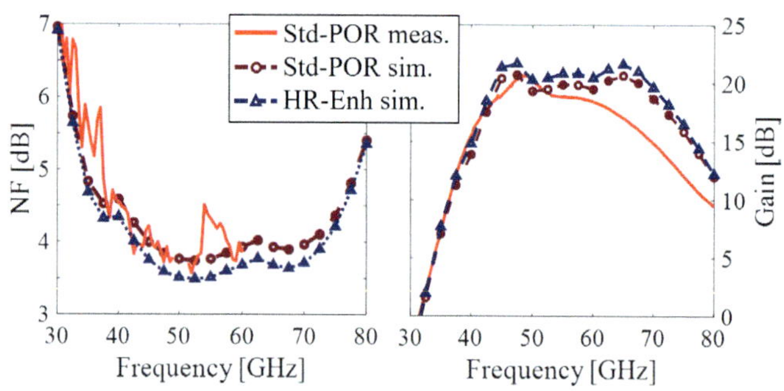

Fig. 9. Measured and simulated performance – noise-figure and gain – of the designed LNA fabricated on the different substrate options.

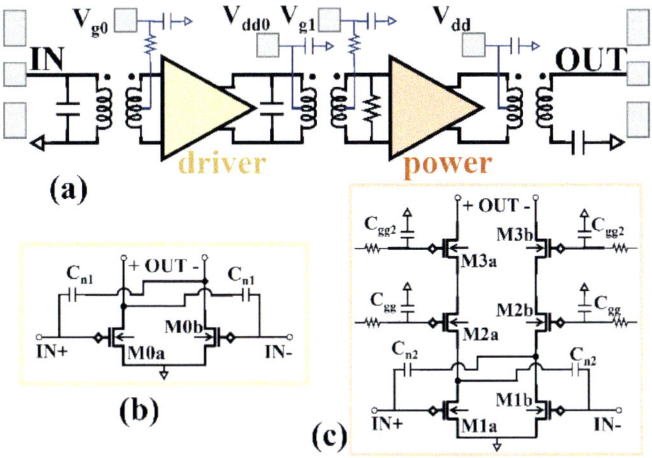

Fig. 10. Schematic of PA1 (a), the driver stage (b) and power stage (c).

the required impedance of the driver stage for optimum large-signal performance. The gate and drain biases are fed to the transistors through the transformers' center-taps as shown in Figure 10a. Large bias resistors are added in the gate bias lines to improve stability in common-mode operation [19]. The back-gate terminals of all FETs are connected to DC pads via large bias resistors to decouple RF and DC operation (not shown in Figure 10).

Power Stage Design

The output stage uses nMOSFET super-low threshold voltage with a gate length of 20 nm and a total FET width of 160 μm for all FETs. The 20 nm gate length is a good trade-off between high RF performance (cutoff, f_t, and maximum oscillation frequency, f_{max}) and reliability, for which the breakdown voltage is not significantly lower than other sub-30-nm gate lengths [1]. The total FET width results in a trade-off between high P_{out} and input impedance transformation over a large bandwidth. Layout parasitics are reduced with staggered-type interconnects at the FETs-level to reduce resistive and reactive elements for lower losses, phase mismatch, and ease of in/output impedance matching, while meeting electromigration rules.

The V_{dd} bias voltage is set to 2.4 V, to present a DC V_{ds} of 0.8 V to each FET, the nominal voltage in this technology. V_{gs} = 0.3 V is selected as a trade-off between significant reduction in third harmonic level in medium-to-high power operation (a sweet spot in linearity exists in CMOS-based PAs biased close to class-B operation) and gain that is fundamental at mm-wave frequencies. The back-gate voltage (V_{bg}) changes the threshold voltage (V_{th}), thus drain current and class of operation. V_{bg} = 0 V is fixed in this design (with V_{gs} = 0.3 V), but access to the back-gate terminal of each FET enables compensating for V_{th}-process variation (–240 to +40 mV for V_{bg} = –0.5 to 3 V, respectively).

The driver stage has a similar design to the power stage but features common-source-only FETs with a width of 80 μm (half the power stage) and biased in the same way.

Output Balun Design

Load-pull simulations of the output stage show that an equivalent impedance made of 45.5 Ohm and 74 fF in parallel

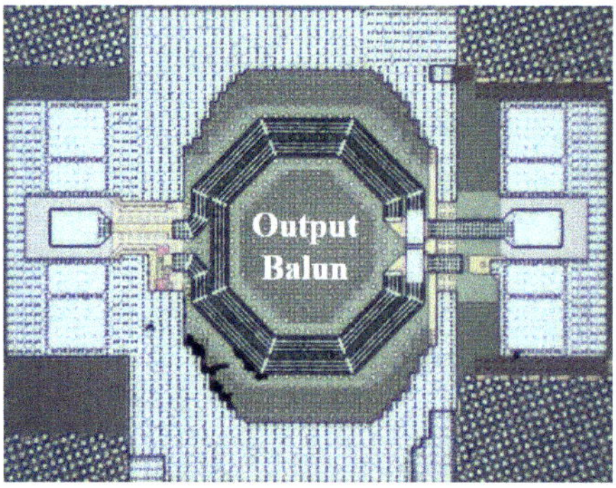

Fig. 11. Microphotograph of the fabricated standalone output balun.

(for each branch), maximizes the output stage large-signal performance over a large bandwidth (optimal load impedance). The output balun is designed based on coupled transmission line elements for an accurate wideband behavior description in a similar way as [16]. The resulting distributed balun is integrated as a standalone structure for thorough characterization, and a microphotograph of that passive device is given in Figure 11.

Figure 12 shows measurements along with initial and post-layout simulations to fit the measurements of the standalone distributed balun. Figure 12b shows the simulated performance of the output matching network as integrated in the PA (including the output RF pad).

An ultra-wideband impedance transformation is well achieved over the whole 24–45 GHz range.

On the Std-POR substrate, this output matching network has an insertion loss of 1–1.5 dB over the band.

Measurements of Substrate Impact on PA Performance

The microphotograph of the full two-stage PA is shown in Figure 13. It has been characterized on-wafer under large-signal continuous-wave measurement conditions at all frequencies from

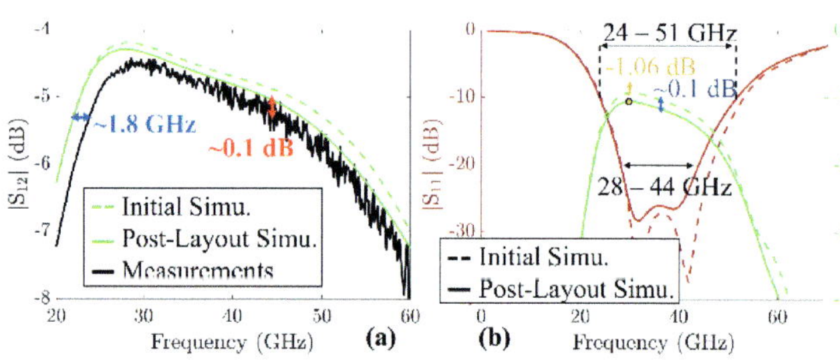

Fig. 12. (a) Simulations (green) and measurements (black) of the standalone output balun on Std-POR. (b) Simulations of the output matching network on Std-POR.

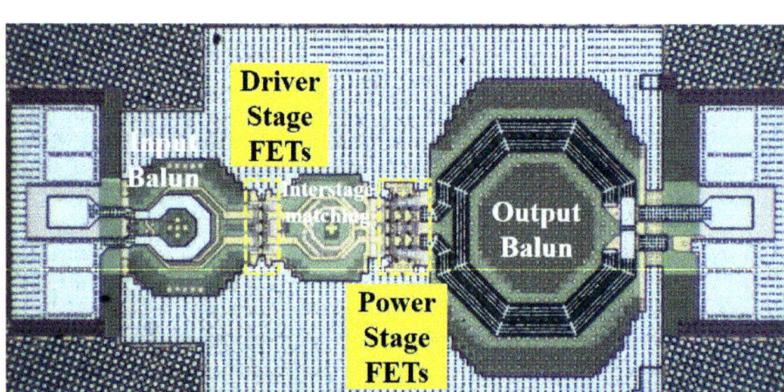

Fig. 13. Microphotograph of the full two-stage PA design.

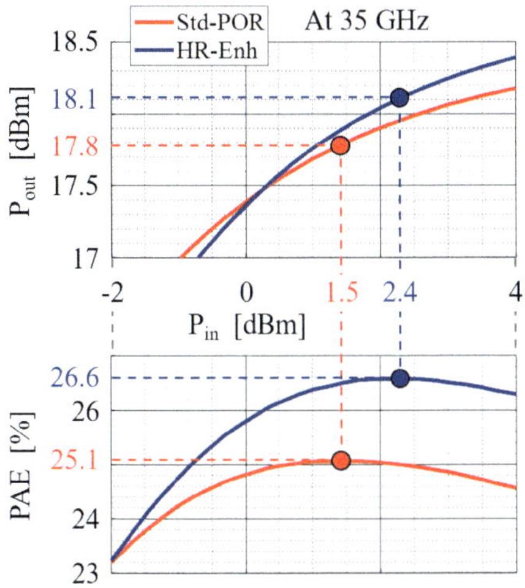

Fig. 14. Measured performance – *Pout* and *PAE* – of the designed PA fabricated on the different substrate options.

24 GHz to 43 GHz, and the results pertaining 35 GHz are plotted in Figure 14. The results demonstrate that by moving the PA design to a higher-resistivity substrate that the peak power added efficiency (PAE) is increased from 25.1% to 26.6%, and that the output power at peak-PAE is increased from 17.8 to 18.1 dBm.

VIII. CONCLUSION

This paper presents the impact of substrate properties on the performance of a wide range of mm-wave passive and active circuits. The devices include CPW lines, spiral inductors, a DC-40 GHz SPDT switch, and 42–67 GHz LNA and a 24–43 GHz PA, all designed and fabricated in GlobalFoundries' 22 nm FD-SOI node on several types of substrates, including both standard (low) and high bulk resistivity wafers, and with several interface passivation options. The main results are recapped in Table 1, and highlight performance gains in the key metrics of each of these devices of the order of 0.3–1.0 dB in the 20–60 GHz range that they cover. By replacing the 10 Ωcm substrate material by a high-resistivity (>500 Ωcm) silicon bulk with appropriate interface passivation, the following gains in device FoMs are demonstrated.

(i) The lineic loss of CPW lines is improved from 1.00 dB/mm (1.26 dB/mm) to 0.15 dB/mm (0.29 dB/mm) at 30 GHz (at 60 GHz).

(ii) The peak quality factor of a 20 GHz spiral inductor is improved from 16.0 to 24.5.

(iii) The insertion loss of a wideband DC-40 GHz SPDT switch module is improved from 2.10 dB to 1.75 dB at 30 GHz.

(iv) The gain and noise figure of a three-stage 42–67 GHz LNA design (cas-

TABLE 1

SUBSTRATE IMPACT ON THE MAIN FIGURES OF MERIT OF THE STUDIED MM-WAVE DEVICES

Device	**Band**	**FoMs**	**Values of FoMs on standard Si**	**Values of FoMs on high-res. Si**
CPW	DC to >100 GHz	Lineic loss at 30 GHz [dB/mm] Lineic loss at 60 GHz [dB/mm]	1.00 1.26	0.15 0.29
Spiral inductor	DC to 20 GHz	Peak quality-factor [/]	16.0	24.5
SPDT	DC to 40 GHz	Insertion loss at 30 GHz [dB]	2.10	1.75
LNA	42 to 67 GHz	Gain at 55 GHz [dB] Noise figure at 55 GHz [dB]	19.90 3.76	20.85 3.51
PA	24 to 43 GHz	Peak PAE at 35 GHz [%] Output power at 35 GHz at peak PAE [dBm]	25.1 17.8	26.6 18.1

code topology with transformer-based gain-boosting loads) are improved from 19.90 dB and 3.76 dB to 20.85 dB and 3.51 dB, respectively, at 55 GHz.

(v) The peak power added efficiency and output power at peak-PAE of a wide-band differential two-stage 24–43 GHz PA are improved from 25.1% and 17.8 dBm to 26.6% and 18.1 dBm, respectively, at 35 GHz.

References

[1] S. N. Ong *et al.*, "A 22nm FDSOI technology optimized for RF/mmWave applications," *2018 IEEE Radio Frequency Integrated Circuits Symposium (RFIC)*, Philadelphia, PA, USA, 2018, pp. 72-75, doi: 10.1109/RFIC.2018.8429035.

[2] M. Rack and J. -P. Raskin, "SOI technologies for RF and millimeter wave applications," *ECS Trans.*, vol. 92, no. 4, pp. 79–94, Jul. 2019, doi: 10.1149/09204.0079ecst.

[3] Y. Wu, S. Gamble, B. M. Armstrong, V. F. Fusco, and J. A. C. Stewart, "SiO2 interface layer effects on microwave loss of high-resistivity CPW line," *IEEE Microw. Guided Wave Lett.*, vol. 9, no. 1, pp. 10–12, Jun. 1999, doi: 10.1109/75.752108.

[4] M. Rack, F. Allibert and J. -P. Raskin, "Modeling of semiconductor substrates for RF applications: Part I—Static and dynamic physics of carriers and traps," *IEEE Transactions on Electron Devices*, vol. 68, no. 9, pp. 4598-4605, Sept. 2021, doi: 10.1109/TED.2021.3096777.

[5] D. Lederer and J. -P. Raskin, "New substrate passivation method dedicated to HR SOI wafer fabrication with increased substrate resistivity," *IEEE Electron Device Lett.*, vol. 26, no. 11, pp. 805–807, Nov. 2005, doi: 10.1109/LED.2005.857730.

[6] F. Allibert *et al.*, "Engineering SOI substrates for RF to mmWave front-ends", *Microwave Journal*, Oct. 2020, online available: https://www.microwavejournal.com/articles/34713-engineering-soi-substrates-for-rf-to-mmwave-front-ends.

[7] M. Rack, L. Nyssens and J.-P. Raskin, "Low-loss Si-substrates enhanced using buried PN junctions for RF applications," *IEEE Electron Device Letters*, vol. 40, no. 5, pp. 690-693, May 2019, doi: 10.1109/LED.2019.2908259.

[8] M. Rack, L. Nyssens and J.-P. Raskin, "Silicon-substrate enhancement technique enabling high quality integrated RF passives," *2019 IEEE MTT-S International Microwave Symposium (IMS)*, 2019, pp. 1295-1298, doi: 10.1109/MWSYM.2019.8701095.

[9] L. Nyssens, M. Rack and J. -P. Raskin, "Effective resistivity extraction of low-loss silicon substrate at millimeter-wave frequencies," *2019 14th European Microwave Integrated Circuits Conference (EuMIC)*, 2019, pp. 1-4, doi: 10.23919/EuMIC.2019.8909575.

[10] W. Heinrich, "Quasi-TEM description of MMIC coplanar lines including conductor-loss effects", *IEEE Transactions on Microwave Theory and Techniques*, vol. 41, no. 1, pp. 45–52, Jan. 1993, ISSN: 1557-9670. DOI: 10.1109/22.210228.

[11] L. Nyssens *et al.*, "PN junctions interface passivation in 22 nm FDSOI for low-loss passives," *2022 24th International Microwave and Radar Conference (MIKON), Gdansk, Poland*, 2022, pp. 1-4, doi: 10.23919/MIKON54314.2022.9924803.

[12] M. Rack, L. Nyssens, Q. Courte, D. Lederer and J.-P. Raskin, "Impact of device shunt loss on DC-80 GHz SPDT in 22 nm FD-SOI," *ESSDERC 2021 - IEEE 51st European Solid-State Device Research Conference (ESSDERC)*, 2021, pp. 195-198, doi: 10.1109/ESSDERC53440.2021.9631835.

[13] M. Rack, L. Nyssens, Q. Courte, D. Lederer and J. -P. Raskin, "A DC-120 GHz SPDT switch based on 22 nm FD-SOI SLVT NFETs with substrate isolation rings towards increased shunt impedance," *2022 IEEE Radio Frequency Integrated Circuits Symposium*

(RFIC), 2022, pp. 83-86, doi: 10.1109/RFIC54546.2022.9863217.

[14] S. Guo, T. Xi, P. Gui, D. Huang, Y. Fan and M. Morgan, "A transformer feedback Gm-boosting technique for gain improvement and noise reduction in mm-wave cascode LNAs," *IEEE Transactions on Microwave Theory and Techniques*, vol. 64, no. 7, pp. 2080-2090, July 2016, doi: 10.1109/TMTT.2016.2564398.

[15] L. Gao, E. Wagner and G. M. Rebeiz, "Design of E- and W-band low-noise amplifiers in 22-nm CMOS FD-SOI," *IEEE Transactions on Microwave Theory and Techniques*, vol. 68, no. 1, pp. 132-143, Jan. 2020, doi: 10.1109/TMTT.2019.2944820.

[16] F. Wang and H. Wang, "24.6 An instantaneously broadband ultra-compact highly linear PA with compensated distributed-balun output network achieving >17.8dBm P1dB and >36.6% PAEP1dB over 24 to 40GHz and continuously supporting 64-/256-QAM 5G NR signals over 24 to 42GHz," *2020 IEEE International Solid- State Circuits Conference - (ISSCC), San Francisco, CA, USA*, 2020, pp. 372-374, doi: 10.1109/ISSCC19947.2020.9063157.

[17] B. Moret, V. Knopik and E. Kerherve, "A 28GHz self-contained power amplifier for 5G applications in 28nm FD-SOI CMOS," *2017 IEEE 8th Latin American Symposium on Circuits & Systems (LASCAS), Bariloche, Argentina*, 2017, pp. 1-4, doi: 10.1109/LASCAS.2017.7948059.

[18] H.-T. Dabag, B. Hanafi, F. Golcuk, A. Agah, J. F. Buckwalter and P. M. Asbeck, "Analysis and design of stacked-FET millimeter-wave power amplifiers," *IEEE Transactions on Microwave Theory and Techniques*, vol. 61, no. 4, pp. 1543-1556, April 2013, doi: 10.1109/TMTT.2013.2247698.

[19] N. Deferm and P. Reynaert, "Differential and common mode stability analysis of differential mm-Wave CMOS amplifiers with capacitive neutralization," *Analog Integr. Circuits Signal Process.*, vol. 80, no. 1, pp. 1–12, 2014, doi: doi/abs/10.5555/2629941.2629967.

[20] B. Park *et al.*, "Highly linear mm-Wave CMOS power amplifier," *IEEE Transactions on Microwave Theory and Techniques*, vol. 64, no. 12, pp. 4535-4544, Dec. 2016, doi: 10.1109/TMTT.2016.2623706.

Martin Rack was born in Oxford, England in 1991. He received his M.Sc. and Ph.D. degrees in electrical engineering from the UCLouvain, Belgium, in 2014 and 2021, respectively. Since then, he has been a senior researcher at UCLouvain working on advanced silicon technologies for RF and mm-wave applications.

To that end, he is involved in the complete modeling of semiconductor substrates along with the development of novel high-performance RF substrate solutions. In particular, he is working on taking substrate effects into account during the design of mm-wave circuit modules in advanced SOI nodes, by developing design flows combining both of full-wave electromagnetic simulation tools and TCAD semiconductor physics solvers. He is also implicated in the development of large-signal models of the non-linear behavior of semiconductor substrate interfaces for the evaluation of substrate-induced harmonic distortion in RF and mm-wave systems. Dr. Rack has published over 75 papers in peer-reviewed journals and conference proceedings.

Lucas Nyssens was born in Ottignies, Belgium in 1993. He received his B.Sc. and M.Sc. degrees in electrical engineering from Université catholique de Louvain, Louvain-la-Neuve, Belgium in 2014 and 2017, respectively. He also received an M.Sc. in electronic engineering from Politecnico di Torino, Turin, Italy in 2017. He completed a Ph.D. degree in electrical engineering at Université catholique de Louvain in 2023. His research focuses on the characterization and modelling of passive structures and MOSFETs on SOI technology at millimeter-wave frequencies towards material properties extraction and device figures of merit assessment. It also includes circuit design to analyze substrate related performance and features at circuit level. His research also involves advanced on-wafer calibration and de-embedding techniques for accurate characterization at millimeter-wave and sub-millimeter-wave frequencies.

Massinissa Nabet received his M.Sc. degree in electrical engineering from the Université des Sciences et de la Technologie Houari Boumediene, Algeria in 2016. Since then, he has been pursuing his Ph.D. degree at Université catholique de Louvain (UCLouvain), Belgium. His current research interests include substrate modeling and high frequency characterization of RF devices in SOI technology. He is also working on circuit design of RF front-end-modules, including mm-wave mixers, switches and power amplifiers.

Dimitri Lederer received M.Sc. and Ph.D. degrees in applied sciences from UCLouvain in 2000 and 2006, respectively. After completing his Ph.D. in the field of SOI material and device characterization for microwave applications he worked in various SMEs and large-scale companies on the development of RF and microwave and sub-THz products. Prior to joining UCLouvain in 2019, he worked as a Senior Engineer at GlobalFoundries, where he worked on the development of the 45RFSOI technology, an Si-based technology targeted for mmWave applications. His research interests cover the modeling, fabrication, design, and characterization of Si-based technologies, devices, and circuits for sub-THz and THz applications. These research activities also include the integration of III–V and more exotic materials on Si. Targeted applications cover imaging, sensing, and telecommunications. Professor Lederer has published more than 100 papers in peer-reviewed journals and conference proceedings.

Jean-Pierre Raskin (IEEE M'97, IEEE SM'06, IEEE F'14) was born in Aye, Belgium in 1971. He received his Industrial Engineer (Ing.) degree from the Institut Supérieur Industriel d'Arlon, Belgium, in 1993, and M.Sc. (Ir.) and Ph.D. degrees in Applied Sciences from the Université catholique de Louvain (UCLouvain), Louvain-la-Neuve, Belgium in 1994 and 1997, respectively. From 1994 to 1997, he was a research engineer at the Microwave Laboratory, UCLouvain, Belgium. He worked on the modeling, characterization and fabrication of MMICs in silicon-on-insulator (SOI) technology for low-power, low-voltage applications. In 1998, he joined the EECS Department of The University of Michigan, Ann Arbor, USA. He has been involved in the development and characterization of micromachining fabrication techniques for microwave and millimeter-wave circuits and microelectromechanical transducers/amplifiers working in harsh environments. In 2000, he joined the Microwave Laboratory of UCLouvain, Louvain-la-Neuve, Belgium as Associate Professor and he has been a Full Professor since 2007. From September 2009 to September 2010, he was visiting professor at Newcastle University, Newcastle Upon Tyne, UK. He was the head of the Electrical Engineering Department of UCLouvain between 2014 and 2017. His research interests are the modeling, wideband characterization and fabrication of advanced SOI MOSFETs as well as micro and nanofabrication of MEMS/NEMS sensors and actuators, including the extraction of intrinsic material properties at nanometer scale. He is an Institute of Electrical and Electronics Engineers (IEEE) Fellow, European Microwave Association (EuMA) Associate Member, Société de l'électricité, de l'électronique et des technologies de l'information et de la communication (SEE) Senior Member, Material Research Society (MRS) Member, International Association of Advanced Materials (IAAM) Fellow, Institution of Engineering and Technology (IET) Member and European Association of Service-Learning in Higher Education (EASHLE) Member. He was the recipient of the Médaille BLONDEL 2015, a famous French reward that honors each year a researcher for outstanding advances in science who has demonstrated a major impact on the electrical and electronics industry. He received the SOI Consortium Award 2016, the European SEMI Award 2017, the Médaille AMPERE 2019, the Georges Vanderlinden Prize 2021, and the IET Achievement Medal in Electronics 2022, in recognition of his vision and pioneering work for RF SOI. He was elected member of the Académie royale de Belgique in 2023. He is author or co-author of more than 350 scientific journal articles. He was the recipient of the European Global Education Innovation Award 2017, Triple E Awards 2022, and UNISERVITATE Award 2022 for the development of the Ethical Service Learning course named IngénieuxSud with the NGO Louvain Cooperation.

Detection of Human Targets Using a MIMO FMCW Radar with Slow-time DC-value Suppression

Keivan Alirezazad, and Linus Maurer

Abstract—**Frequency-modulated continuous wave (FMCW) radars provide distinct movement-based features that aid with remote human target identification. A 2D complex-valued range-angle image (RAI) is produced by applying a 2D fast Fourier transform (FFT) on fast-time samples of virtual receiver array generated by the time-division multiplexing access (TDMA) technique. Unlike the human chest wall, which produces a significant phase shift from one chirp to the next in FMCW radar, stationary clutters exhibit an approximately constant phase during the coherent processing interval (CPI). In this study, an advanced 77-GHz multiple-input-multiple-output (MIMO) FMCW radar is employed, and a DC-value suppression technique is implemented along the slow time axis. This novel approach results in a stationary-clutter-free 2D complex-valued range-angle image (RAI), enabling accurate identification of human targets. Compared to analogous research studies, the main advantage of this approach is its high precision and low computational complexity for real-time applications. The conducted experiments and robust graphical representations demonstrate the applicability of the proposed method in a complex, congested environment.**

Index Terms—**Human target identification, mm-wave radar, MIMO, FMCW, RAI, TDMA, CPI.**

This research was conducted as part of the BEYOND5 project, which was jointly funded by the Horizon 2020-ECSEL JU under grant agreement number 876124 and the German Ministry of Education and Research (BMBF) under project number 16MEE0004. The authors are also grateful for the assistance provided by the DTEC Vital Sense project. The European Union is funding dtec.bw through the NextGenerationEU program.

K. Alirezazad and L. Maurer are with the Institute for Microelectronics and Circuit Technology of the Universität der Bundeswehr München-Neubiberg, Germany (email: keivan.alirezazad@unibw.de, linus.maurer@unibw.de).

I. INTRODUCTION

THE automotive market is currently the primary target of mm-wave radar applications. However, various possible industrial and therapeutic applications are being actively researched. In recent years, there has been a growing interest in a new area of research known as human vital signs detection using mm-wave radars. This technology can be useful in a variety of scenarios, from locating buried people in catastrophic accidents to monitoring driver fatigue and drowsiness to avoid fatal vehicle collisions [1]. The first processing stage in implementing this application is human target identification. When the precise position of human subjects is necessary, this use case can be expanded to smart home infrastructure and indoor monitoring. The developed method should be able to suppress moving and stationary clutters in the background to achieve this goal.

The standard deviation [2], [3] and the exponential moving average (EMA) algorithm [4], derived from the moving target indicator (MTI), are the two most frequently employed techniques for suppressing stationary targets while preserving moving ones.

Authors in [5] introduce a new approach derived from EMA for suppressing stationary and undesired moving clutter sources indoors. On the other hand, paper [6] suggests a curve-length method for identifying human subjects in the experimental scene by estimating the length of the I/Q signal trajectory.

In this paper, a novel method is proposed to suppress the stationary clutters for indoor human identification utilizing a 77-GHz multiple-input-multiple-output (MIMO) frequency modulated continuous wave (FMCW) radar. We accomplish this by monitoring the phase shift of the radar's raw data along the slow time axis.

In contrast to stationary clutters, the human chest wall, an experimental sample of vibrating sources, yields a significant phase shift from one chirp to the subsequent one in FMCW radars due to respiration and heartbeat. Consequently, the use of a slow-time DC-value suppression technique significantly attenuates stationary clutters, thereby facilitating the localization of human subjects. This method's low computational complexity and high performance make it an attractive alternative to the previously presented schemes. To the authors' best knowledge, this is the first time that the feasibility of applying a slow-time DC-value suppression method to radar's range-angle image (RAI) for human target recognition is investigated.

The remaining sections of this paper are organized as: Section 2 is devoted to the fundamentals of MIMO FMCW radars. Section 3 provides a thorough theoretical explanation of the proposed method. Section 4 demonstrates the effectiveness of the proposed method through clear visualizations and presents the empirical results derived from a radar board operating at the central frequency of 77 GHz. Finally, Section 5 summarizes the paper.

II. FMCW RADAR SIGNAL PROCESSING: THEORETICAL FOUNDATIONS

Time-division multiple access (TDMA) is a concept used by MIMO FMCW radar systems to expand the number of receiving channels and to compute the Azimuth angle with higher precision. After employing the TDMA technique and generating a series of uniformly spaced chirps, various operations, such as down-conversion, filtering, and digitization, are applied to the received signal in parallel channels to extract the so-called beat signal [7]. Figure 1 depicts our proposed approach for human recognition. Each component of this diagram will be thoroughly discussed in the sections that follow.

The mathematical form of an FMCW radar's beat signal $g_b(t)$ is as follows [8]:

$$g_b(t) \cong \frac{1}{2}\cos\left(2\pi f_c \gamma + 2\pi\alpha\gamma t - \pi\alpha\gamma^2\right)$$
$$\gamma = \frac{2R(\tau)}{c} = \frac{2\left[R_0 + \xi(\tau)\right]}{c} \tag{1}$$

where f_c denotes the carrier frequency of the chirp signal, γ is the round-trip delay of the reflected chirp signal, t stands for the fast-time and c is the speed of light. $R(\tau)$ represents the evolution of range along slow-time τ, and the $\xi(\tau)$ denotes the tiny time-variant displacement of the chest wall induced by cyclic contraction and relaxation of the cardiac muscles and respiratory activity. In addition, the chirp rate α is defined as the ratio of the chirp bandwidth B to the chirp time T_{UP}.

In our case, α is a factor of 10^{13}Hz/s and γ is a factor of 10^{-8} s; hence, $\alpha\gamma^2 << 1$ and so

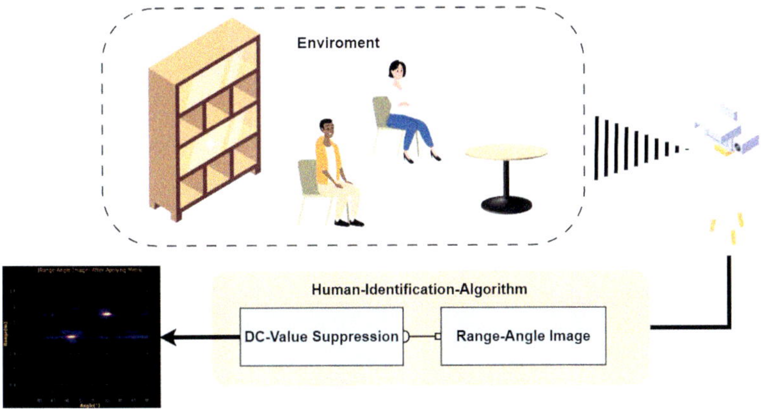

Fig. 1. Block diagram of the proposed human identification algorithm using MIMO FMCW radar.

the third term in (1) may be disregarded. As a result, the simplified form of (1) would be:

$$g_b(t) \cong \frac{1}{2}\cos\left(\phi(\tau) + 2\pi f_b t\right)$$
$$\phi(\tau) = \frac{4\pi f_c[R_0 + \xi(\tau)]}{c},\ f_b = \frac{2\alpha R_0}{c} \tag{2}$$

where $\phi(\tau)$ is the phase history that is analyzed using several chirps to determine the chest wall displacement, and f_b denotes the beat frequency. Since the chest wall remains almost constant during a single chirp, the effect of $\xi(\tau)$ on the beat frequency f_b has been ignored.

III. PROPOSED METHOD

Our proposed method consists of (a) generation of a range angle image (RAI) using MIMO FMCW radar, (b) DC-value compensation of a IQ signal.

Creating an RAI

Given that a single chirp includes N fast-time samples and the number of synthesized virtual antenna elements is L, RAI generation consists of the following steps [7]:

1) The beat signal of a chirp signal is captured for each antenna element and formatted as a 2D matrix of size $N \times L$, with each column containing the fast-time samples of a unique antenna element.
2) A P-point fast Fourier transform (FFT) is applied to this 2D matrix along each column to create a range-profile matrix that is $P \times L$ in size. Each row of the range-profile matrix corresponds to a distinct range bin and holds spatial characteristics of the objects in the radar's field of view (FoV) at that range.
3) A Q-point FFT is performed on each row of the range profile matrix to generate a range-angle spectrum of size $P \times Q$. Prior to FFT in fast-time and angular dimensions, the zero-padding technique was used to decrease the effects of discretization and improve FFT resolution. As a result, $P > N$ and $Q > L$.

Since FMCW radar transmits a series of chirp signals known as frames, the preceding procedure is repeated for all the constituent frames of a coherent processing interval (CPI), and the resultant RAIs are arranged as a 3D matrix Ψ that is $M \times P \times Q$

in size, where $M = U \times W$. The symbols U and W denote the number of chirps in a single frame and the total number of frames that constitute the CPI, respectively.

Slow-time DC-value Suppression in MIMO FMCW Radar

As derived in the preceding section, Ψ is a matrix with real I and imaginary Q components. Thus $\Psi[m, p,q]$ can be expressed as follows [9]:

$$\Psi[m,p,q] = I_{ac}[m,p,q] + I_{dc}[p,q] + j(Q_{ac}[m,p,q] + Q_{dc}[p,q]) \quad (3)$$

with $m = 1,\ldots, M$, $p = 1,\ldots, P$, and $q = 1,\ldots,Q$. In (3), I_{dc} and Q_{dc} denote the DC values of the 2D Fourier transform's real and imaginary parts, respectively. In contrast, I_{ac} and Q_{ac} represent the sinusoidal components arising from phase shifts due to infinitesimal displacements in a given range and angle bins. By suppressing the DC values, the estimate

$$I_{ac}[m,p,q] = \Re(\Psi[m,p,q]) - \underbrace{\frac{1}{M}\sum_{m=1}^{M} \Re(\Psi[m,p,q])}_{I_{dc}[p,q]} \quad (4)$$

of the real part's sinusoidal component and the estimate

$$Q_{ac}[m,p,q] = \Im(\Psi[m,p,q]) - \underbrace{\frac{1}{M}\sum_{m=1}^{M} \Im(\Psi[m,p,q])}_{Q_{dc}[p,q]} \quad (5)$$

of the imaginary part's sinusoidal component are computed (here, $\Re(z)$ and $\Im(z)$ denote the real part and the imaginary part of z, respectively). In the end, a stationary-clutter-free heatmap $\Psi'[m, p, q]$ given by

$$\Psi'[m,p,q] = I_{ac}[m,p,q] + jQ_{ac}[m,p,q] \quad (6)$$

is generated to facilitate human identification within the radar's FoV.

IV. EXPERIMENT SETUP

Utilized Hardware

This study uses Radarbook2 (RBK2) [10], a 77-GHz MIMO FMCW radar produced by INRAS GmbH, that communicates with the host computer through TCP/IP and has a maximum data rate of 550MBit/s to avoid loss of data. The clock signal of ADCs in the RF front-end supports both 40 MHz and 20 MHz. The radar uses a 1-D antenna array topology. It consists of two transmit and 16 receive antennas with an inter-element spacing of 0.5 λ and 7.5 λ respectively, where λ is the freespace wavelength at 77 GHz. Hence, the radar synthesizes 30 non-overlapping and two overlapping virtual components by employing the TDMA technique. In the primary measurement mode, the timing unit of the radar, **SeqTrig**, which runs in the FPGA part of the system, creates a trigger signal for the RF synthesizer to begin the preset chirp sequence and for the ADC data interface, **DatInt**, to capture the data during the up-chirp.

Configuration Parameters

Table I lists the configuration parameters of RBK2 for this research project. Considering the parameters of this table and assuming c as the speed of light, the range and Azimuth angle bins of the generated RAI, Δr and $\Delta\theta$, are respectively calculated as follows:

$$\Delta r = \frac{cT_{\mathbf{UP}}}{2B} \cdot \frac{f_{\mathbf{s}}}{P} = 1.87\ cm \quad (7)$$

$$\Delta\theta = \frac{2}{L} \cdot \frac{180}{\pi} = 1.79° \quad (8)$$

TABLE I

SIMULATION PARAMETERS OF THE RADAR-BASED HUMAN IDENTIFICATION ALGORITHM

Parameter	Description	Value	Unit
B	Bandwidth	4	GHz
f_c	Central Frequency	75	GHz
f_s	Fast Time Sampling Frequency	2	MHz
f_{FRR}	Slow Time Frame Repetition Rate	12.5	Hz
T_{UP}	Up-chirp Duration	128	μs
N	Number of Samples per each Chirp	256	–
L	Number of Virtual Antennas	31	–
U	Number of Chirps in a Frame	1	–
W	Number of Frames in CPI	50	–
P	Range-FFT Size	512	–
Q	Angle-FFT Size	256	–

Experimental Procedure and Validation of Results

The experimental setting utilized to demonstrate the effectiveness of the proposed approach is depicted in Figure 2. Two test subjects participated in this experiment; one was sitting on a chair at (2.1(m), 20°) and the other at (2.8(m), –35°). Furthermore, at the laboratory, two corner cubes were positioned in front of RBK2 to provide significant static clutter. The radar and corner cubes were mounted on tripods. The human subjects' positions in the generated RAI are obscured because the power of the transmitted chirps is reflected more properly off these pyramidal metallic objects than off the other elements in the scene, making human identification more challenging.

Fig. 2. Data acquisition setup with two human subjects. The radar and corner cubes are both placed on tripods.

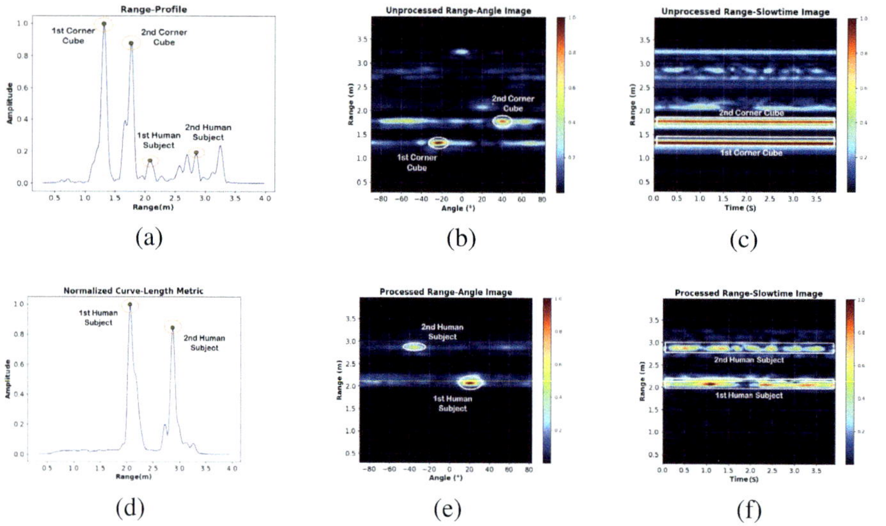

Fig. 3. Empirical results of the test scenario involving two humans: (a) the range-profile derived from one of the receiving antennas, marking the positions of the test participants and corner cubes; (b) the unprocessed RAI with indistinct coordinates of the human subjects; (c) the unprocessed time-evolving range-profile matrix; (d) the normalized amplitude of 1D curve-length metric; (e) the processed RAI in which the first and second participants are detected at (2.1(m), 20°) and (2.8(m), –35°), respectively; and (f) the time-evolving range-profile matrix after employing the DC-value suppression technique.

Figure 3 displays the results of the test scenario. Figures 3a-3d show the raw radar data, while Figures 3e and 3f show the processed radar data, which has undergone the DC-value suppression technique. The time-evolving range-profile matrix has a steady peak at the range bins occupied by the corner cubes, as illustrated in Figure 3c. Nonetheless, the range bins where the human test subjects are positioned show significant amplitude fluctuations owing to respiratory activity (Figure 3f).

A 2D matrix is populated with the fast-time samples of the third virtual antenna across CPI. Following that, we validated the range bins of the identified test participants by applying the 1D version of the curve-length metric described in [6] to this matrix. The existence of two peaks at 2.1(m) and 2.8(m) in Figure 3d correlates to the presence of two horizontal stripes at the identical range bins in the processed time-evolving range-profile matrix (Figure 3f). Moreover, compared to the second human subject, the peak associated with the first human subject has a higher amplitude in Figure 3d. This point is consistent with a more notable reflection of this individual in the processed RAI (Fig. 3e).

Figure 4 demonstrates the effect of DC-value suppression in this particular case in the form of a trajectory plot. This graphic illustrates the imaginary and real values of the processed RAI over CPI at two distinct positions. The magenta markers corresponds to the range and angle bins in which the first test participant is positioned, while the green point is related to one of the used corner cubes. As can be seen, the sinusoidal phase variation of the blue trajectory line with magenta markers indicates the existence of a non-stationary object. Conversely, the green point remains still at the center of the coordinate system throughout the entire CPI.

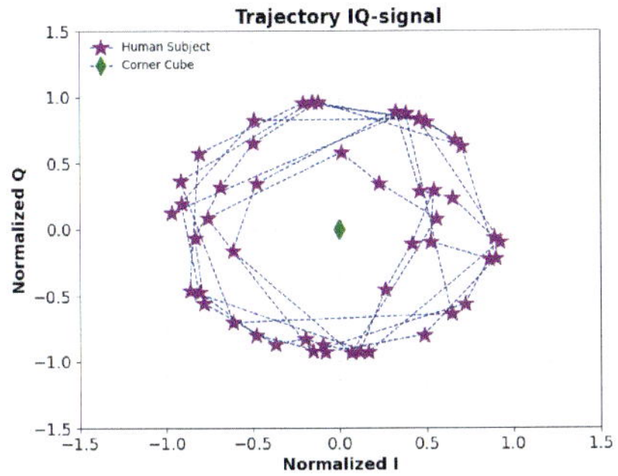

Fig. 4. Trajectory plot of the normalized IQ-values over CPI (4*s*) after employing the DC-value suppression method. The magenta markers correspond to the test participant's coordinates, while the green point relates to one of the corner cubes utilized.

V. Conclusion

This paper proposes a DC-value suppression technique along the slow-time axis of a 77-GHz multiple-input-multiple-output (MIMO) frequency-modulated continuous wave (FMCW) radar. The technique is aimed at suppressing stationary clutters for indoor human identification. The developed experimental scenario and comparisons demonstrate the practicality of the proposed method in complex, cluttered scenarios for MIMO FMCW radars.

References

[1] A. Ahmad, J. C. Roh, D. Wang and A. Dubey, "Vital signs monitoring of multiple people using a FMCW millimeter-wave sensor," 2018 IEEE Radar Conference (RadarConf18), Oklahoma City, OK, USA, 2018, pp. 1450-1455, doi: 10.1109/RADAR.2018.8378778.

[2] Y. -H. Shen, Y. -R. Chien and S. -H. Fang, "Poster abstract: Human detection with weak ranging signal for FMCW radar systems," 2020 19th ACM/IEEE International Conference on Information Processing in Sensor Networks (IPSN), Sydney, NSW, Australia, 2020, pp. 343-344, doi: 10.1109/IPSN48710.2020.00-15.

[3] Z. Peng et al., "A portable FMCW interferometry radar with programmable low-IF architecture for localization, ISAR imaging, and vital sign tracking," IEEE Transactions on Microwave Theory and Techniques, vol. 65, no. 4, pp. 1334-1344, April 2017, doi: 10.1109/TMTT.2016.2633352.

[4] C. Will, P. Vaishnav, A. Chakraborty and A. Santra, "Human target detection, tracking, and classification using 24-GHz FMCW radar," IEEE Sensors Journal, vol. 19, no. 17, pp. 7283-7299, 1 Sept.1, 2019, doi: 10.1109/JSEN.2019.2914365.

[5] P. Nallabolu, L. Zhang, H. Hong and C. Li, "Human presence sensing and gesture recognition for smart home applications with moving and stationary clutter suppression using a 60-GHz digital beamforming FMCW radar," IEEE Access, vol. 9, pp. 72857-72866, 2021, doi: 10.1109/ACCESS.2021.3080655.

[6] K. Han and S. Hong, "Detection and localization of multiple humans based on curve length of I/Q signal trajectory using MIMO FMCW radar," IEEE Microwave and Wireless Components Letters, vol. 31, no. 4, pp. 413-416, April 2021, doi: 10.1109/LMWC.2021.3057867.

[7] K. Alirezazad, G. Rhiel and L. Maurer, "2D CNN-GRU model for multi-hand gesture recognition system using FMCW radar," 2022 20th IEEE Interregional NEWCAS Conference (NEWCAS), Quebec City, QC, Canada, 2022, pp. 158-162, doi: 10.1109/NEWCAS52662.2022.9841948.

[8] J. Gamba, Radar Signal Processing for Autonomous Driving, 1st edn., Singapore: Springer Nature Singapore Pte Ltd., 2020, pp. 45-51.

[9] M. Alizadeh, G. Shaker, J. C. M. D. Almeida, P. P. Morita and S. Safavi-Naeini, "Remote monitoring of human vital signs using mmWave FMCW radar," IEEE Access, vol. 7, pp. 54958-54968, 2019, doi: 10.1109/ACCESS.2019.2912956.

[10] "RBK2-77G-TX2RX16-D01 frontend," INRAS GmbH, Linz, Austria, April, 2019.

Keivan Alirezazad received his B.Sc. degree in control engineering with a specialization in electrical engineering in 2012. He went on to obtain his M.Sc. degree in electrical engineering, majoring in information technology, from the University of Rostock in Germany in 2019. In 2020, he joined the Institute for Microelectronics and Circuit Design at the University of the Bundeswehr in Neubiberg, Germany, where he is currently pursuing a Ph.D. degree. His research interests encompass various fields, including radar signal processing, audio signal processing, computer vision, machine learning for automotive high-resolution MIMO radar systems, speech and audio coding technology, and biomedical applications. His dedication and contributions to the field have been recognized, and he was honored with the 24th International Microwave and Radar Conference (MIKON) young scientist award in 2022.

Linus Maurer (SM, 14) received the Diploma Engineer degree in Physics and his Dr. techn. degree from the Johannes Kepler University Linz, Austria, in 1997 and 2001, respectively. In 2002 Prof. Maurer joined DICE, an Infineon Technologies design center dedicated to the development of cellular RF-transceivers and ICs for automotive radar applications. Between 2007 and 2012 he was site-manager for the Sense&-Control division of DICE overlooking the Automotive Radar development activities. Since August 2012 Prof. Maurer is full Professor for "Electronic Components and Integrated Circuits" at the University of the Armed Forces in Neubiberg, Germany. He was "IEEE Distinguished Microwave Lecturer" from 2007 to 2009, received the "MTT-S Outstanding Young Engineer Award" in 2010 and is recipient of the "ITG-Förderpreis" 2002 and the EEEfCOM-price 2002 and 2006, respectively. His main research interests are focused on wireless communication and mm-wave radar systems. He has authored and co-authored over 160 publications in these fields and has given numerous international presentations and tutorials.

Appendix

Speaker name: **Aarno Pärssinen**
Organization name: **University of Oulu**
Country: **Finland**
Profile picture:

Presentation title: **Some visions towards 6G cellular systems**

SHORT ABSTRACT OF PRESENTATION

The first major 6G program in the world was established at the University of Oulu in 2018. The first visions have been followed by many projects in different continents, including European flagships Hexa-X and Hexa-X-II, to enhance different enabling technologies from hardware and signal processing to future concepts. The presentation will discuss how these visions have evolved and elaborate some of the expectations for RF and hardware technologies and circuit design to facilitate advances in 6G.

A new generation of cellular systems has been established roughly once per decade. Change in generation allows larger changes and improvements than other releases to the evolution of cellular communications leading the advances in the wireless and ICT industry. After 5G matured to a stable standard, the vision of the next mobile communication generation (6G) started to develop. While 4G and 5G already provide a large diversity of different radios and use cases, two distinct fields in technology evolution were addressed: (1) radios for seemingly infinite capacity and extremely long life and (2) low power for IoT. This presentation focuses mostly on the first field.

Initial visions and somewhat wishful thinking of 1 Tbps data rates at 1 THz carrier frequencies has been grounded with early system/technology analysis and early prototypes. Still, in 2023 there is some time to mature thoughts before actual standardization begins. At the same, a lot of debate is ongoing in various projects on the real need of specific high-end applications for the data rate and latency. At the same time, the need for improved capacity in the market continues with existing networks. As always, it is difficult or impossible to predict the future and to identify the killer apps before they exist. Therefore, matching vision work, estimates of the real needs of final users of the technology, and technological capability to produce mature products must be considered both separately and as joint opportunities for the industry.

From a hardware perspective, semiconductor and other RF technologies start to be on the edge of being capable of meeting future needs. As evolution of RFIC and packaging technologies are not following the same evolutionary path with the same speed as digital, new thinking in architectures, used technologies and packaging in addition to pushing the boundaries of affordable semiconductor technologies are needed in order facilitate this development. Combining possible future needs of users and applications to the required radio network and link capacity considering physics-based constraints of hardware technologies is not a straightforward task. When working towards 6G, it is of major importance to be hardware aware or even hardware friendly when designing radio protocols and use of spectrum to achieve the ambitious targets.

Early prototypes from ICs to testbeds accomplished at the University of Oulu with collaborators are complementing the theoretical observations presented in this talk.

KEYWORDS

6G mobile communication, 3GPP, integrated circuits, phased arrays, physical layer, radio communication, radio frequency, radio transceivers, RFIC

ACKNOWLEDGEMENT

This work was supported in part by the Academy of Finland, 6G Flagship program under Grant 346208, in part by the Business Finland Project RF Sampo under Grant 3071/31/2021 and in part by the European Commission through the H2020 Project Hexa-X under Grant 101015956. GlobalFoundries is acknowledged for providing silicon fabrication through the university program and Keysight for measurement equipment.

SHORT BIOGRAPHY SPEAKER

Aarno Pärssinen received his D.Sc. degree from the Helsinki University of Technology, Finland, in 2000. From 2000 to 2011 he was with Nokia Research Center, Helsinki, Finland where he served as a member of CEO Technology Council from 2009 to 2011. From 2011 to 2014, he was at Renesas Mobile and at Broadcom. Since September 2014 he has been with University of Oulu, Centre for Wireless Communications, Finland where he is currently a Professor. He leads the devices and circuits research area in the 6G flagship program. His research interests include transceiver architectures and RFICs for wireless systems. He has authored and co-authored one book, two book chapters, more than 200 international journal and conference papers, and holds several patents. He was a member of the technical program committee of ISSCC in 2007–2017, chairing the wireless subcommittee in 2014–2017.

Speaker name: Didier Belot
Organization name: STMicroelectronics
Country: France
Profile picture:

Presentation title: 6G communications and localization: An overview of technologies opportunities and challenges

SHORT ABSTRACT OF PRESENTATION

Telecom communities are beginning the preparation of the next generation mobile communications (6G) and are presenting KPIs towards Tbits/s, 300 GHz carrier frequencies, space multiplexing, accurate localization, etc. To serve these challenges, microelectronic roadmaps must be revised.

This presentation presents a benchmark of existing RF and millimeter wave solid-state transceiver technologies. It zooms in on the power amplifier and low noise amplifier functions as they are the most demanding building blocks in solid-state RF design. RF process properties are assessed in view of the upcoming specifications and requirements.

This presentation also explores representative use cases of beyond 5G and 6G and identifies which process is the most suited for which function and application. This analysis covers RF-CMOS, SiGe HBT, and the more exotic III-V HBT and HEMT. Their assets, capabilities, and limitations to support mm-wave and sub-THz applications are discussed while considering their cost efficiency and their suitability for co-integration and packaging.

In conclusion, this presentation proposes different transceiver partitioning architectures that might serve 6G applications operating at the higher frequency bands by the end of this decade and proposes microelectronics process trends that might satisfy these applications.

KEYWORDS

6G, B5G, technology roadmap, RF-CMOS, SiGe, III-V, power amplifier, low noise amplifier

SHORT BIOGRAPHY SPEAKER

Didier first spent 30 years in ST technology R&D, where he worked initially on characterization modeling of bipolar transistors before moving to high frequency analog design to develop optical–electronic interfaces. In 1995, he moved to analog-RF design to work on cellular and Bluetooth transceivers, in SiGe and CMOS respectively. In 2006, he created a joint team with LETI, to initiate the development of mmW R&D prototypes at 60GHz in CMOS-65, which led to a demonstration of a 4Gbs wireless link at 60GHz over 1m.

In 2014, he joined CEA-Leti, to continue his research work, on mmW CMOS, on mm-wave plastic guides, and on very high speed mm-wave (1Gbs → 10GBs and more). He has also played an important role in the orientation of the RF and mm-wave roadmaps and in the management of programs such as IPCEI within LETI.

At the beginning of 2023 he came back to ST in the Technology Design Platform group, in charge of Wireless strategy and Innovation. Moreover, he has participated, and is participating in, the elaboration of European and worldwide Roadmaps on Wireless, (NEREID, CORENECT, ECS-SRIA, IRDS ...). His current research interests include mm-wave propagation through plastic, sub-THz communications, III–V devices on silicon for mm-wave and THz applications, and the use of RF for quantum computing. He was a member of the French National Scientific Council "Micro and Nanotechnologies" from 2012 to 2016, and a member of various conference program committees such as RFIC, ESSCIRC, ISSCC, and IEDM. He is a member of the IEEE-MTT-9 (mmW & THz Devices to System) technology committee and the EuMW technical committee. He is a reviewer for IEEE MTT and SSC journals, and is author or co-author of more than 400 publications and 70 patents.

Speaker name: Piet Wambacq
Organization name: imec
Country: Belgium
Profile picture:

Presentation title: **Heterogeneous integration for complex mm-wave transceivers**

SHORT ABSTRACT OF PRESENTATION

The need for more bandwidth in wireless communication drives carrier frequencies towards the mm-wave frequency region, where beamforming with antenna arrays is needed to obtain a sufficient link distance. Similarly, radar is moving out to higher frequencies and larger bandwidths for better range, speed, and angle resolution. Both radar and communication transceivers are very complex systems, requiring advanced CMOS for integration of most of the functionality. Only in the front-end is CMOS falling short, especially when the transmit power at mm-wave frequencies needs to be generated at a sufficiently high efficiency. An extra complication for the transmit part in wireless communication systems with complex modulation schemes, is that the power amplifier needs to operate in its linear regime. This regime implies its operating point to attain a backoff from of maximum output power. To overcome the power limitations of CMOS, other semiconductor technologies such as SiGe BiCMOS and compound semiconductors (such as InP and GaN) can yield higher energy efficiency.

The combination of CMOS with non-CMOS front-end circuits and an antenna array requires heterogeneous integration. During this talk, we consider such a communication system to operate around 140 GHz. This frequency is in the middle of the D-band, which will be (partially) used for the highest data rate applications foreseen in 6G.

When using beamforming, the spacing between the antenna elements (e.g. patches) needs to be a half wavelength $\lambda/2$ in free space (i.e. 1.1 mm at 140 GHz) to concentrate all energy in the main lobe of the array's radiation pattern. This puts a constraint on the footprint of the transceiver chips that are connected to the patches. Small antenna arrays can be designed as a linear array. In this case, the pitch requirement only puts a constraint on the chip size in one dimension. For the integration of the antenna array with the CMOS part and the non-CMOS front-end, a technology such as silicon interposer can be used. The electronics can be mounted on the carrier substrate with low-loss, wideband transitions between the chips and the carrier. The routing to the antenna elements can be designed with striplines.

In larger antenna arrays, the patches need to be positioned in a two-dimensional matrix fashion. As a result, the pitch of the front-end electronics needs to obey the λ/2 pitch in both directions. To keep the wiring between the front-end electronics and the antenna array simple, a 3D stack can be used comprising the antenna array and the front-end circuitry that is placed right underneath. If the beamforming electronics is fabricated on a different substrate from the front-end electronics, it could be placed underneath the latter.

Another challenge is the design of the active circuits. These need to be wideband to accommodate a few adjacent RF channels corresponding to multiple users. Of course, the availability of beamforming enables spatial multiplexing, giving an outlook to even more users. Another design challenge is the power amplifier (PA). For a given EIRP, the number of antenna patches, and hence the number of PAs, can be traded for the output power per transmit path. A small number of PAs relaxes the required output power per PA and hence the design challenge and the potential thermal problems. On the other hand, many antenna patches require more phase shifters and a splitting of a single signal over more antenna paths, causing extra power consumption. Clearly, there is an optimum number of antenna elements. Further, when comparing different semiconductor technologies in this aspect, it turns out that this optimum number of antenna elements is lower and comes with an overall lower power consumption for InP thanks to its higher efficiency. In any case, for any semiconductor technology, designing transceivers for wireless communication at carrier frequencies beyond 100 GHz with data rates of several tens of Gbits per second, is challenging and the design must be tackled in conjunction with the design of the antennas and the packaging.

All these system integration and technology choices will be presented and traded-off to identify possible solutions for future mobile systems.

KEYWORDS

6G, mobile communication, CMOS, III–V, InP, antennas, integration, packaging, heterogeneous integration, beamforming

Short biography speaker

Piet Wambacq (Senior Member, IEEE) received his M.Sc. degree in electrical engineering and Ph.D. degree from Katholieke Universiteit Leuven, Leuven, Belgium, in 1986 and 1996, respectively. In 1996, he joined imec, Leuven, where he is currently Fellow, working on analog/RF/mm-wave IC design in various technologies for wireless applications. Since 2000, he has been a Professor with Vrije Universiteit Brussel (VUB), Brussels, Belgium. He has authored or co-authored six books and more than 350 articles in edited books, international journals and conferences. He was a member of the Program Committee of ISSCC from 2012 to 2020 with the role of RF Subcommittee Chair from 2016 to 2020. He is the chair of the program committee of ISSCC 2023. He has been a member of the Program Committee of the ESSCIRC and DATE conferences. He was an Associate Editor for IEEE Transactions on Circuits And Systems—Part 1 from 2002 to 2004 and a Distinguished Lecturer of the IEEE Solid-State Circuits Society of IEEE from 2016 to 2018.

Speaker name: Thomas Kämpfe
Organization name: Fraunhofer IPMS
Country: Germany
Profile picture:

Presentation title: D-Band noise characterization and modelling in advanced FDSOI devices

SHORT ABSTRACT OF PRESENTATION

Noise parameters allow for the precise prediction of the noise figure of active devices, such as low-noise amplifiers, which are required for communication or radar applications. Recent developments to increase the carrier frequency for communication to the mmWave frequency range, increases the need for improved noise parameter characterization and modeling in advanced nodes. Therefore, fully depleted silicon-on-insulator (FDSOI) is a promising technology for ultra-low power communication and radar applications. In this work, we will discuss the analysis and modeling of high-frequency noise parameters in advanced 22 nm FDSOI CMOS devices up to the D-band (110–170 GHz) frequency regime.

Specifically, we will present experimental results of multi-finger n-channel thin-oxide (0.8 V) MOSFETs with 18 nm gate length and 0.5 μm finger width in common-source configuration. The S-parameters and the noise parameters of the devices were characterized by using dedicated tuner-based noise measurement setups in three different frequency bands, 8–50 GHz, 75–110 GHz (W band), and 110–170 GHz (D band). A probe-tip calibration was performed on an impedance standard substrate by employing the enhanced line-reflect-reflect-match calibration algorithm. The small-signal parameters such as transconductance or the gate capacitances are extracted from the multi-bias S-parameters. To simplify the analysis, the substrate is neglected, and the access resistances are considered constant.

The noise characteristics of the devices are predicted using the validated Pucel (PRC) noise model. The noise of the intrinsic FET is modeled by two correlated noise current sources, the induced gate noise and the channel thermal noise. The result from the analysis of the model parameters shows that the noise correlation is necessary to achieve good model accuracy.

KEYWORDS

FDSOI, D-band, noise modeling, PRC

SHORT BIOGRAPHY SPEAKER

Thomas Kämpfe works as a group manager for CMOS integrated RF & AI at Fraunhofer IPMS and deputy department manager at Center Nanoelectronic Technologies. He received his habilitation in electrical engineering in 2022, his Ph.D. degree in Physics in 2016 and his Diplom degree in Physics in 2011, all from TU Dresden. After research visiting scholar positions with the University of Colorado at Boulder and Stanford University, he joined the Fraunhofer Society in 2017. To date, Dr. Kämpfe has authored and co-authored more than 160 peer-reviewed articles in top journals and conferences such as Nature Scientific Reports, Advanced Functional Materials, IEEE TC, IEDM, VLSI, ESSCIRC, and DATE. His current research interests include brain-inspired computing, approximation computing, and computing-in-memory. He obtained the Dresden Excellence Award in 2023, the Excellent Paper Award at RFIT 2022 and Best Paper Nomination in DATE 2021. He has served on technical programs, organization committees and paper review for many international conferences such as ICICDT and AICAS as well as journals such as TCAS-I, TCAD, and TETCSI.

Index

16 QAM 52, 57–9, 61, 63, 65, 67–8
3D Printing 87, 91
3GPP 3, 19, 21–2, 28–9, 52, 116, 152
5G 1–6, 8–17, 19–24, 26, 28–9, 31, 33, 50–2, 54–5, 57, 67, 69–74, 76, 78, 80, 82–4, 87, 88–90, 92–4, 96, 98–100, 103–5, 107–15, 121, 123, 141, 153
6G 1–2, 4, 6, 8–10, 15–6, 28, 67, 71, 103, 151–6
Antenna Array 7, 23, 26, 29, 31–2, 50, 60, 70, 72–3, 87, 91, 93–4, 98, 100, 146, 155–6
Antennas 3, 7, 26, 32, 46, 50, 60, 70, 73, 83, 87–8, 90–1, 95, 98–9, 100–1, 146, 148, 156

Bandpass Filter (BPF) 31, 45
Base Station (BS) 22–4, 31, 33, 35, 124
Beam Control 87–8, 93, 98
Beamforming 7, 8, 17, 20, 31–2, 35, 46, 50–1, 55, 69, 70, 72, 87–9, 98, 149, 155–6
Bit Error Rate (BER) 46, 65, 67

Capacitive Digital-To-Analog Converter (CDAC) 37–9
Channel Aggregation 55, 57–8, 60, 69
CMOS 7–9, 13, 16, 25, 29, 33, 41, 51–5, 57, 67–9, 87, 98, 103–6, 108–14, 116, 118, 123–5, 127, 137, 141, 153–8
Coplanar Waveguides (CPW) 127
Coherent Processing Interval (CPI) 143, 145–9

D-band 9, 55, 57–65, 67–9, 155, 157–8
DC-Offset Compensation (DCOC) 37, 118
Differential Non-Linearity (DNL) 43
Digital Base-Band (DBB) 26–7, 32, 35, 63, 65, 72
Digital Beamforming (DBF) 31–2, 34, 50
Digital Pre-Distortion (DPD) 26, 31, 35–6, 41, 50, 106–9, 115–7, 119, 123
Direction-Of-Arrival Estimation 87, 89, 98
Doherty PA (DPA) 103, 106–110
Dynamic Element Matching (DEM) 42

Effective Number Of Bits (ENOB) 24, 26, 39
Efficiency 4, 8, 9, 1–5, 19, 35, 39, 44, 46, 50, 67, 72, 87–8, 93, 103–16, 121–3, 127–9, 139–40, 153, 155–6
Electro-Magnetic (EM) 46, 74
Electro-Static Discharge (ESD) 35–6, 44
Engineered Substrates 1
Envelope Tracking (ET) 103, 105, 111
Electronically Reconfigurable Superstrate (ERES) Antenna 87, 91–3, 99
Electronically Steerable Parasitic Array Radiator (ESPAR) Antenna 87–9, 99

Fully Depleted Silicon On Insulator (FDSOI) 1, 7, 9, 11–5, 26, 28, 31, 33, 39, 40, 43, 50, 52–4, 69, 111, 127–30, 134, 139–41, 157–8
Flip Chip Assembly 71
Frequency-Modulated Continuous Wave (FMCW) 143–6, 148–50
Fan-Out Wafer-Level Packaging (FOWLP) 103, 108
Frequency Generation 57, 67
Frequency Synthesis (Sx) 26, 41, 44–6

Heterogeneous Integration 155–6
High-Data Rate Wireless 57, 67
High-Resistivity Substrate 127, 129
Human Target Identification 143

III-V 8, 10, 16, 114, 116, 142, 153–4, 156
Inductors 108, 127–9, 132, 134–5, 139
InP 1, 9, 10–1, 15–6, 155–6
Input-Referred Third-Order Intercept (IIP3) 24, 28, 35–7, 52
Integral Non-Linearity (INL) 43
Integrated Circuit (IC) 16, 25, 51, 67–9, 83, 90, 110–13, 140, 150

Integration 5, 7–9, 12–5, 17, 19, 21, 29, 31, 33, 50, 68, 70–1, 73–4, 76, 82–4, 89, 91, 94–5, 103–4, 110, 142, 153, 155–6
Internet of Things (IoT) 11, 14, 17, 53, 69, 87, 99

Laterally-Diffused Metal-Oxide
Semiconductor (LDMOS) 103–6, 110
Least Significant Bit (LSB) 38, 43
Lens Antenna 61, 87–8, 90, 99, 100
Linear Feedback Shift Register (LFSR) 43
Low NoiseAmplifier (LNA) 4, 28, 31, 35–7, 52, 116, 127–9, 134–6, 139
Local Oscillator (LO) 25–7, 32, 35–6, 39, 41, 44, 47–8, 50, 59, 62–3, 67–8, 116
Low Drop Out Voltage Regulator (LDO) 40
Low Pass Filter (LPF) 31
Low Power 11–2, 15, 19–20, 22, 31–2, 35, 40, 51, 54–5, 69, 87, 98, 114, 124, 134, 142, 157

Milimeter Wave (mm-wave, mmW) 11–5, 31,
45, 52–3, 55, 71–3, 84–5, 127–9, 137, 139,
141–3, 150, 153–6
Mobile Communication 15–6, 99, 152–3, 156
Most Significant Bit (MSB) 37–8, 43
Multiband 103–4, 108, 110, 112–3
Multimode 103–4, 108–10, 113
Multiple Input Multiple Output (MIMO) 3, 15, 19, 21, 23–5, 29, 31–2, 46, 50–1, 72, 87–8, 94, 98, 100, 115, 143–6, 148–50

Noise Figure (NF) 35, 46, 128

Orthogonal Frequency-Division Multiplexing
(OFDM) 88

Packaging 27, 71, 73–4, 76–7, 82–5, 113–4, 152–3, 156
Periodically Repeated Oscillations Train (PROT) 50
Phased Arrays 88, 99, 152
Physical Layer 70, 152
Pilot Line 11, 14
Piezo-On-Insulator (POI) 1, 6, 9–10
Poly Phase Filter (PPF) 45
Power Amplifier (PA) 9–10, 25, 29, 31–2, 35, 40, 42, 103–16, 121–3, 127–9, 136, 138–41, 156
Power-Added Efficiency (PAE) 128
Pucel (PRC) 157–8
Printed Circuit Board (PCB) 44, 73, 84, 95
Process Voltage & Temperature (PVT) 42, 46
Proportional To Absolute Temperature Current Reference (IPTAT) 40

Quadrature Amplitude Modulation (QAM) 21, 24, 47, 51–2, 57–9, 61, 63, 65, 67–8, 108, 121, 141
Quadrature Phase Shift Keying (QPSK) 35, 108

Radar 11, 13, 15, 53, 143–50, 155, 157
Radio Communication 8, 67, 152
Radio Frequency (RF) 11, 15, 51, 55, 67–8, 83, 87, 93, 111–3, 140, 152
Radio-Frequency Integrated Circuit (RFIC) 31
Range-Angle Image (RAI) 143–8
Receiver (RX) 35, 58
Reconfigurable Antenna 87, 94–5, 99
RF Design 55, 71, 153–4
Radio-Frequency Front-End (RFFE) 1–3, 13–4, 31–2, 113, 116, 142, 146
RF Switches 92, 127
RF-CMOS 33, 127, 153
Radio Frequency Integrated Circuit (RFIC) 25–7, 31–3, 50–2, 54–5, 67–8, 72–3, 75–6, 78, 80, 82, 93, 111–3, 123, 125, 140–1, 152, 154
Radio Frequency Silicon On Insulator (RFSOI) 1, 4–5, 9, 11–7, 57, 60, 68, 106, 111

Serial Peripheral Interface (SPI) 31, 39
Silicon Germanium (SiGe) 9, 33, 103–4, 109–10, 113, 153–5
Signal to Noise and Distortion Ratio (SNDR) 39
Silicon On Insulator (SOI) 1, 4–7, 9–14, 16–7, 28, 31, 52–5, 69, 103–14, 116, 123, 127–8, 140–2

Smart Cut 11–2, 15–6
Spurious-Free Dynamic Range (SFDR) 24, 52, 39, 119
Successive Approximation Analog-to-Digital Converter (SAR ADC) 31, 37–9
Super Low Vt (SLVT) 41
Supply Modulator 111–2, 115–6, 120–1, 123
Switched-Beam Antenna 87, 99

Time Devision Multiple Access (TDMA) 143–4, 146
Technology Roadmap 1, 153
Thermal Management 71, 74–5
Teraherz (THz) 9, 11, 15–6, 20, 67, 84–5, 142, 153–4
Transceiver (TRX) 19, 25, 27, 35, 46, 52, 54, 57–9, 67, 72, 114, 123, 125, 150, 152, 154–6
Transmitter (Tx) 58

User Equipment (UE) 7, 35, 52, 72

Value Chain 11–4, 16
Vehicle-To-Everything (V2X) 89, 115–6
Voltage Digital-to-Analog Converter (VDAC) 46

Wireless Sensor Networks 83, 87